T0247858

ALSO BY SIMON PARKIN

The Island of Extraordinary Captives: A Painter, a Poet, an Heiress, and a Spy in a World War II British Internment Camp

A Game of Birds and Wolves: The Ingenious Young Women Whose Secret Board Game Helped Win World War II

Death by Video Game: Danger, Pleasure, and Obsession on the Virtual Frontline

THE
FORBIDDEN
GARDEN

THE BOTANISTS OF BESIEGED LENINGRAD AND
THEIR IMPOSSIBLE CHOICE

SIMON PARKIN

SCRIBNER
New York   London   Toronto   Sydney   New Delhi

Simon & Schuster: Celebrating 100 Years of Publishing in 2024

For information about special discounts for bulk purchases, please contact Simon & Schuster Special Sales at 1-866-506-1949 or business@simonandschuster.com.

The Simon & Schuster Speakers Bureau can bring authors to your live event. For more information or to book an event, contact the Simon & Schuster Speakers Bureau at 1-866-248-3049 or visit our website at www.simonspeakers.com.

Interior design by Erika R. Genova
Maps © Mike Parson / Barking Dog Art 2024

Manufactured in the United States of America

1 3 5 7 9 10 8 6 4 2

Library of Congress Control Number: 2024030974

ISBN 978-1-6680-0766-2
ISBN 978-1-6680-0768-6 (ebook)

Image Credits
Pages 2, 10, 154, 202, 276, and 312–19 courtesy of VIR; pages 40 and 70 Boris Kudoyarov/Sputnik via AP; page 82 courtesy of Getty Images; page 96 via Wikimedia Commons; page 136 RIA Novosti Archive/Boris Kudoyarov via Wikimedia Commons; page 182 Michael Traham/Sputnik via AP; page 218 Scientific Archive of the Russian Academy of Arts. Collection of Gasilov S.G.; page 232 courtesy of University of Jena Archives via Wikimedia Commons; page 260 United States Holocaust Memorial Museum; page 292 ©RBG Kew

For Huey and Jed.
May you forever know peace,
And daily have bread.

We shall go to the pyre. We shall burn.
But we shall not retreat from our convictions.

—Nikolai Vavilov, March 1939

If a brother or sister is lacking in daily food,
and one of you says to them, "Go in peace, be warmed and filled,"
without giving them the things needed for the body,
what good is that?

—James 2:15–17

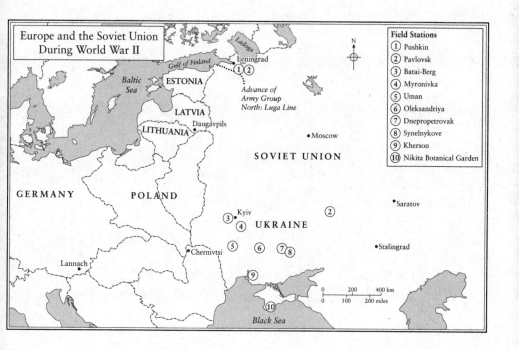

Europe and the Soviet Union During World War II

Field Stations
1. Pushkin
2. Pavlovsk
3. Batai-Berg
4. Myronivka
5. Uman
6. Oleksandriya
7. Dnepropetrovak
8. Synelnykove
9. Kherson
10. Nikita Botanical Garden

Ladoga
Gulf of Finland
Leningrad
① ②
Baltic Sea
ESTONIA
LATVIA
Daugavpils
LITHUANIA
Advance of Army Group North: Luga Line
Moscow
SOVIET UNION
GERMANY
POLAND
Kyiv
③•
④
UKRAINE
Saratov
②
⑤ ⑥ ⑦⑧
Chernivtsi
Stalingrad
Lannach
⑨
⑩
Black Sea

0   200   400 km
0   100   200 miles

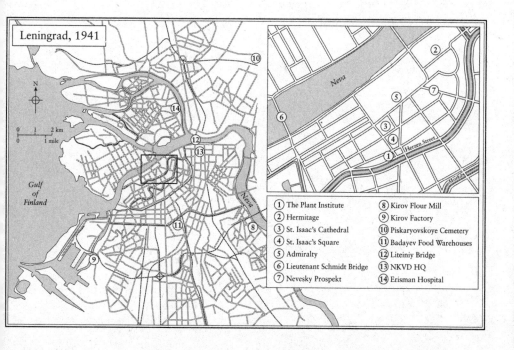

Leningrad, 1941

Neva
N
Gulf of Finland
Neva
Herzen Street
Moika

1. The Plant Institute
2. Hermitage
3. St. Isaac's Cathedral
4. St. Isaac's Square
5. Admiralty
6. Lieutenant Schmidt Bridge
7. Nevesky Prospekt
8. Kirov Flour Mill
9. Kirov Factory
10. Piskaryovskoye Cemetery
11. Badayev Food Warehouses
12. Liteiniy Bridge
13. NKVD HQ
14. Erisman Hospital

0   1   2 km
0   1 mile

CONTENTS

AUTHOR'S NOTE

The siege of Leningrad—the Russian city now and formerly known as St. Petersburg—lasted for almost nine hundred days. It was the longest blockade in recorded human history. By conservative estimates the siege claimed the lives of approximately three-quarters of a million people, four times the number that died in the atomic bombings of Nagasaki and Hiroshima combined. In 1942 every third person living in the city perished. Most died from starvation, or "dystrophy" as the Soviet authorities euphemistically referred to the cause of death. *The Forbidden Garden* contains vivid descriptions of the effects of extreme malnourishment on the human body and psyche, up to and including death.

This is a work of historical narrative nonfiction. The events described in this book are not fabricated but drawn from diaries, letters, memoirs, oral histories, newspaper reports, and other primary source material according to the best recollections of the various players and protagonists. Quotations and dialogue recalled by any speaker or witness years after the event in question should be taken as impressionistic rather than verbatim. Where there are discrepancies in dialogue between sources, the version closest to the date of the event described has been used.

For clarity, the Latin names for plants, especially wild varieties and weeds identified by the botanists to feed Leningrad's citizens in the summer of 1942, have been included, alongside the common Russian names used at the time. A full list of references and sources can be found at the end of the text.

Nikolai Vavilov (*center*) sits among colleagues and students at Saratov University's
Research Laboratory in January 1921. Two months later Vavilov and twenty-seven
of his colleagues would travel to Petrograd—soon to become Leningrad—
and develop the world's first seed bank.

PROLOGUE

PETROGRAD
MARCH 1921

THE TRAIN DREW INTO PETROGRAD Station and nodded to a standstill. A thirty-three-year-old man flanked by twenty or so young scientists disembarked. It had been a cold and arduous journey from Moscow. Springtime approached, but the train had rattled through a landscape still dressed for winter. No horizon marked where the snow-shrouded hills met the frigid sky. Passengers gazed out of the window into an eternal freeze-frame. Out there, only the occasional whisper of wind interrupted the white silence, and the frosted branches of the birch trees dipped and swayed like players in a spectral ballet. On the platform the young men and women stretched, yawned, and unkinked their backs.

Nikolai Vavilov felt at ease out there, in the expanse of nature, where he had spent much of his life as a botanist and explorer. Still, it felt good to be in the city, a long-anticipated arrival to this world-renowned center of arts and science. Thickset and tall, Vavilov had a striking appearance. When exploring the peaks and deserts of Iran, or the mountain plateaus of Tajikistan, he often wore a three-piece suit, white collar, tie, and felt fedora. His dark hair was swept back from his high forehead to frame angular features that gave him a sculptural quality. Lean and muscular, he cut an impressive figure atop a hill, surveying the landscape. Here on the platform a passerby might, from a closer vantage, notice the calloused hands, roughened from a career of fieldwork, an intimate sign of his deep and enduring connection to the earth.

Vavilov began to organize his young colleagues. Some were former students, whom he had taught at the provincial university in Saratov, where the week's journey began—testament to his powers as a teacher, and his magnetism as a leader. He was not a man who allowed the character of his interactions to be unduly influenced by hierarchy and status. He addressed chancellor and student alike with warmth and clarity, a luminary to whom people were drawn to follow. Together the group began to unload heavy boxes of scientific books, papers, and equipment, and bags and crates filled with a collection of rare seeds that Vavilov had been keeping for this day.

The atmosphere was one of keen anticipation. Vavilov, a bright young star in the firmament of Russian science, had recently been appointed as director of the Bureau of Applied Botany and Plant Breeding, a botanical institute dedicated to the study of flora and plant life, founded in the city in 1894. He had come to Petrograd armed not only with books, tools, and samples, but also the kind of ambitious plan to which young people are willing to pledge their lives: a dream to turn the city's penniless Plant Institute[*] into the world's first seed bank, a facility to store and preserve seeds for future use in agriculture, research, and conservation. The Institute, as he had described it to a friend a few months earlier, would be nothing less than "a treasury of *all known* crops and plants."

To furnish his vision, Vavilov had recently begun to mount overseas expeditions, hunting for and collecting rare seeds to bring to Russia. The specimens currently stowed in the group's baggage would combine with those already held at the Institute to form the basis of a new and historic collection. As soon as his team of explorer-botanists

---

[*] Originally known as the Institute of Applied Botany and New Crops, then from 1930 as the All Union Institute of Plant Breeding, and since 1992 as the Vsesoyuzny Institut Rastenievodstva (VIR). For ease and elegance throughout this book the VIR is referred to simply as the Plant Institute.

were settled in, they would again head into the wilderness, visiting the farthest reaches of the planet to harvest seeds from even its most remote nooks. They would then return here, to the center of Russia's most storied and magnificent city. Next, Vavilov and his team would begin to experiment, using the seeds to develop hybrids from plants with different genetic qualities, evolving new varieties able to withstand different climates. This would be the second step in his grand plan: to breed supercrops with the capacity to end human famine.

It was a lofty aim with desperate relevance: Russia was currently gripped by nationwide famine, and the recent world war had segued into a civil conflict that had crippled the country's food production and distribution systems and isolated its people from international aid. War had compounded and highlighted the shortcomings of the Russian economy. Inflation, profiteering, the collapse of food supplies, and the accelerated breakdown of authority had stirred up discontent. Strikes and looting began a crescendo of unrest, culminating in a political coup that had transferred power from the tsars to the Bolshevik usurpers, led by the Revolutionary leader Vladimir Lenin.

The Bolsheviks had nationalized private property and created a so-called Provisioning Army, which roamed Russia's villages and robbed the peasants of their food stores to distribute them among urban populations. Those who resisted, or who were found to have hidden grain or other supplies, were subject to an immediate death sentence. This state-endorsed banditry led to civil war. Drought and crop failures exacerbated these human-made problems. As the young botanists collected their belongings from the train, across the country hundreds of thousands of ordinary people were now starving to death.

Outside the station, Vavilov arranged for the group's supplies to be loaded onto horse-drawn carts. His staff expressed astonishment at their leader's ability to conjure small miracles. Then the group stepped into a beleaguered city. After his coup Lenin had departed for Moscow, leaving Petrograd in a squall of hunger, political tension, and ideological conflict. If the Revolution had been the boozed-up party, the city had now plunged into the bruised aftermath. Amenities sputtered and services crashed; many homes were without heat or water. Once known as the cradle of transformation, Petrograd—which would soon adopt the name of its departed leader to become Leningrad—had become a graveyard. Across a city snared in the tectonic changes of history, the torn, faded banners of revolution whipped in the wind.

Vavilov had known the city was beset by hardship when he accepted his new role. But the potential to realize his vision for the world's first seed bank outweighed the risks. Besides, as he put it in a letter to a friend, "man does not live by bread alone." Still, he had not counted on the scale of the troubles its population faced. Petrograd was "dying before my very eyes," recorded the artist Mstislav Dobuzhinsky, who tried to capture the city's "terrible, deserted, and wounded look" in a series of lithographs that year. The month Vavilov arrived, the Russian writer Maxim Gorky reported to his friend the English writer H. G. Wells that the people had almost run out of food: "It can be said without exaggeration that there will soon be famine in Petrograd."

Vavilov and his followers traipsed along the city's main thoroughfare, Nevsky Prospekt, through a scene of melancholic grandeur. On the train the botanists had joked that the Bolsheviks, in their eagerness to rename places and monuments, might perhaps rechristen this main street, as a tribute to Vavilov's botanical interests: the Avenue of Homologous Lines. Now, the baleful reality struck. Gray sullen survivors roamed everywhere, their faces angled downward, their backs drooped like question marks that appeared to ask "What now?" Lines of hungry citizens awaited their rations of moldy bread crusts and frozen offal. "Westward the sun is dropping," wrote the poet Anna Akhmatova. And "already death is chalking doors with crosses, and calling the ravens, and the ravens are in flight."

The group neared the Neva River. Elegant bridges spanned its glinting tributaries, connecting the city's fragmented districts. They crossed to Vasilyevsky Island and finally arrived at the Plant Institute's offices, which were currently situated on the Makarova embankment, housed in a grand six-story-tall building with an internal courtyard. A few years earlier the building's owner had installed a 150-seat cinema inside, complete with a public meeting hall, a stage, and a foyer. From the outside, at least, the Institute appeared to be in fine shape.

Still, Vavilov was apprehensive. His predecessor, Robert Regel, had recently died of typhus while awaiting a train to the provinces, where he had planned to rendezvous with his family. Before he died, Regel had written a letter warning Vavilov of the Institute's impoverished state. A fine and principled scientist, Regel had managed as best he could with what little he was given. During his tenure, he had written letters to scientists across Russia, pleading for them to send him packets of local seeds for study. By the time he fled the city he had amassed a collection of nearly fourteen thousand varieties of wheat, barley, oat, rye, and other seeds and plants—the makings of a robust, scientifically meaningful collection. But, Regel warned, the building and its staff had been ravaged by years of underfunding.

As they entered the building, nothing had prepared Vavilov and his team for the decrepitude they found inside. The group felt the cold billow through the corridors. The central-heating pipes had burst, and the temperatures had dropped below freezing. As they moved from room to room, it appeared as though the offices had been ransacked by a particularly thoughtless invader. Layers of dirt and dust covered scientific equipment and papers. Chairs and desks had been chopped up and used for firewood, leaving the rooms barren of furniture. The few lonely remaining technicians were dejected and lacking in leadership; there was no sense of purpose or duty, no feeling of being part of a greater scheme.

Most excruciating of all to Vavilov, the storage tins in which Regel's seeds had been contained lay open and empty. The specimens had been eaten, either by famished staff or looters. It was, as one of the group recorded in their diary, "a picture of almost complete destruction." The botanists had inherited the corpse of an abandoned enterprise. As Vavilov put it in a letter to a friend that week, "There are a million problems."

And yet, as he stood amid this scene of desecration, inside this starving city in which he had pledged to make his new life—and convinced others to follow—the botanist perceived flickers of hope and opportunity. Perhaps here was the chance to build something new and better from the wreckage. In that moment Vavilov vowed that he would find new premises, a building with space and heating and furniture, and room for laboratory experiments. And he would find land in the suburbs for plots and greenhouses where he and his acolytes might test the fruits of their research. The task was immense but Vavilov was ready to fight.

No matter how long it might take to repair the damage and build up the vision, Vavilov and his staff resolved that never again would the Plant Institute fall into disrepair. And if their great ambitions should come to fruition, and together they succeeded in establishing the first living library of seeds amassed in human history, each man and woman vowed they would preserve this trove no matter the cost.

PART ONE

Nikolai Vavilov stands in his office at the Plant Institute in 1930, surrounded
by maps and samples of wheat, grain, and other seeds collected from recent
expeditions. That year Vavilov visited Great Britain and the United States, and the
Royal Horticultural Society welcomed him as an honorary member.

I

AN EXPLORER VANISHES

UKRAINE AUGUST 1940
NINTEEN YEARS LATER

I

ON AUGUST 6, 1940, AS sunlight knifed through the clouds overhead, Nikolai Vavilov and some of his colleagues gathered outside a youth hostel in the southwestern Ukrainian town of Chernivtsi. On the far side of the continent, Adolf Hitler's troops had overrun France and were poised to invade Britain, where aerial dogfights busied the skies above the home counties. The botanists spared little thought to the foreigners' war this morning, however. The atmosphere was one of high-spirited adventure, a mood set by the group's eminent and ebullient fifty-two-year-old leader.

Few things thrilled Vavilov so much as an expedition to an unfamiliar region. He was a natural pioneer. "If you have ten rubles in your pocket, travel," he often urged his friends. His instinctive excitement today was compounded by a sense of mission: a quest to find and collect new varieties of plants to be returned to the seed bank, which during the past twenty years had become renowned throughout the world. These trips were to Vavilov as panning for gold is to prospectors. Each day the same thrilling proposition: What blank spaces in the collection might be filled by sundown?

Soon after he and his young team had arrived in the city now known as Leningrad, two decades earlier, Vavilov had secured new, larger premises—a three-story nineteenth-century tsarist palace grand enough to house the world's first seed bank. Designed by the architect Ivan Yefimov, it was enviably placed at 42 and 44 Herzen Street, just off St. Isaac's Square, in the center of the city. From this base Vavilov and his staff began to travel the world seeking rare seeds, tubers, roots, and bulbs, to bring to the seed bank to be sorted, cataloged, and stored.

The mission was urgent. Everywhere conflict, natural disaster, and the destruction of habitat threatened to make certain types of plants extinct. Once destroyed, these specimens and their unique characteristics would be irretrievably lost; no amount of genetic tinkering could bring them back. The extinction of unexamined plant varieties could mean the loss of world-changing medicines, or super-crops that could enable communities and nations to protect themselves against famine.

The idea of a seed bank was novel, and the long-term value of a repository of genetic plant material had yet to be fully understood. Some viewed Vavilov's project as an eccentric waste of time and money. Undeterred, Vavilov and his team had, within a decade of their arrival, made great gains in their mission, replacing the initial collection that had been eaten by their predecessors during the Revolution years, and substantially adding to it. By 1933 the botanists had collected at least 148,000 live seeds and tubers, their work motivated by the repetitious waves of famine in Russia, which provided a clear and harrowing link between the theoretical scientific work and its practical ambition to establish food security.

The seed bank had become world-famous. As a journalist for the *Times* of London wrote that same year, it was a "living museum . . . unrivalled in completeness by any other collection in the world." Scientists had started to refer to the project simply as "the world collection of plants." The Plant Institute became a matchless library of the planet's flora that contained enough latent life to—once planted, grown, and harvested—feed every citizen within Leningrad, the Soviet Union, and elsewhere besides.

By the time he arrived in Ukraine, Vavilov had led 115 excursions into 64 countries and countless cities, from Ethiopia to Kazakhstan, Tokyo to Los Angeles, the Sierra Madre to the Silk Road. Today's trip to the Carpathian Mountains was closer to home than many of Vavilov's more exotic sojourns, but no less rich with potential. He believed that several thousand years ago early explorers had carried with them wheat plants from northeast Africa to Europe. Perhaps, he reasoned, descendants of the plants that once fed the people of Egypt and Babylon during the time of the first pharaohs had survived somewhere within the mountain range. Any ancient wheat plants that had endured centuries of cold air in the northern steppes could be ideal candidates for Vavilov's current mission: to breed supercrops able to withstand inhospitable climes in the northern regions of the Soviet Union.

Outside the hostel, Vavilov readied his empty backpack, which he hoped to fill with seeds and specimens before nightfall.

II

EXPERIENCE HAD NOT DULLED VAVILOV'S childlike sense of wonder. As his friend the eminent English biologist Professor Cyril Darlington once put it, "Wherever he went, Vavilov took sunshine and courage." He was charismatic and magnetic, described by the American Nobel laureate Hermann Muller, "more life-loving, life-giving and life-building than anyone else I have ever known." Nevertheless, the past twelve months had been trying. International fame and status had pushed Vavilov unwillingly into the shadow world of Stalin-era politics. Stress had reduced his health. One colleague noted that the "old sparkle" had left the botanist's eyes, along with his "usual slightly ironic cheerfulness." The doorman at the Plant Institute noticed how he became short of breath whenever he climbed the building's grand staircase.

"It's my heart, my dear fellow," Vavilov confided.

Counter to his genial mien, Vavilov had become increasingly prone to fits of rage, which burned out quickly, leaving him feeling awkward and embarrassed. "The brakes are getting worn down," he muttered to his staff, by way of apology. By the time he reached Chernivtsi, Vavilov had amassed a collection and, with houses in Moscow, Leningrad, and Pushkin, a life that was the envy of botanists around the world. Yet it was partly the jealousy of his peers that had needled at his health and led him on several occasions to attempt to resign from his position as director of the seed bank. To be free from his sniping enemies for a few days, doing the work he loved, gave him more pleasure than he had experienced in some time, an infectious lightness of spirit that had spread throughout the group, on this warm, promising morning.

It helped that Vavilov was among the people he loved: botanists and scientists he had known, in many cases, since they were his own students. Vavilov attracted faithful devotees. In conversation, he offered the full and flattering beam of his attention. His "bright, intelligent eyes" made even the most fleeting of encounters unforgettable. At the Plant Institute he had collected a staff of keen, dedicated individuals, each one as committed to their leader as they were to his vision. "He reminds me very much of Mozart," Gavriil Zaitsev, director of one of the Plant Institute's large experimental stations in Central Asia, wrote to his wife. "He is very companionable and straightforward, in spite of his worldliness."

Vavilov's pride at the seed bank was braided through with the affection he felt for his colleagues, whom he referred to as the "kings and queens" of their various specialties. Unlike Stalin and his followers, Vavilov took no interest in a person's background, whether they came from "pure" peasant stock or a more well-heeled background. After all, the circumstances of one's birth were random. Instead, he examined a person's character and sought to nurture shoots of talent. In 1931 he expressed his admiration for the reports written by a colleague who ran the Institute's Polar field station, urging his former student to commit more of his thoughts and observations to paper.

You're such a dummy for not writing more. You have just the right
style, and you know a fair amount, too. You must write, you must put
your ideas into words, you must break through the ice. . . . You should
spend three hours a day writing, just as you eat and drink every day.
. . . Soon they'll be watching you from both the North Pole and the
South Pole.

Few of Vavilov's crew were so dedicated as Vadim Lekhnovich, a
plant taxonomist who, thirteen years earlier, had found in the Institute
both a vocation and—when he met his colleague Olga Voskresenskaya—a partner and wife. Vavilov had sought out Lekhnovich, a young
man with a bushy beard who was serving on the city council at the
time. On their first meeting he urged Lekhnovich to study potatoes,
describing them as "only second to bread" in importance to humankind. Lekhnovich agreed and joined Vavilov's team as a trainee, quickly
working his way from laboratory assistant to his current role of senior
researcher, in which he had become first an expert in the Jerusalem
artichoke, and then in potatoes.

For the trip to Ukraine, Lekhnovich had been happy to share
a room with Vavilov. Like most of his colleagues, he viewed the
seed bank's director as a kind of swashbuckling saint. That morning
Lekhnovich had awoken to find his boss seated at a writing desk by
the door, scribbling notes for the day ahead. A polyglot who read
the eighteenth-century botanist Carl Linnaeus in the original Latin,
Vavilov possessed seemingly inexhaustible energies. On expeditions
he slept for only a few hours at night and routinely worked eighteen-
hour days. He had, as one colleague wrote, "a mind that never slept
and a body which for its capacity for enduring physical hardships
can seldom have been matched." "Life is short," he often said. "One
must hurry." But Lekhnovich had detected a new sense of urgency
to his leader's demeanor on this trip, the agitation of a person wor-
ried that he might not have enough time to achieve his remaining
ambitions.

For the day's expedition Vavilov planned to travel with the head of the Land Department in the lead car. The group would drive the hundred or so miles into the Carpathian Mountains, near the village of Putyla. But even with three vehicles there was not enough room to take everybody who wanted to accompany Vavilov. Some of the Institute staff would have to stay behind to make room for local guests. Lekhnovich was one of the fortunate ones. He would accompany an associate professor from a local university in one of the other two vehicles, "an old car with worn tires" belonging to the people's commissar of agriculture of Ukraine.

As those left behind waved them off, the convoy departed. The cars began to wend their way through the foothills. The journey would normally take three hours, but Lekhnovich knew that, with Vavilov in the lead, it could easily take much longer. While driving through unfamiliar terrain, the seed bank's founder would press his nose to the window and survey the plants, crops, and trees in the fields that lined the roads. Often, he would order the driver to stop the car, take a cutting from a nearby plant, and place it in a small, damp cloth bag. In this mode of specimen gathering, there was little chance to keep to the schedule.

Before long Lekhnovich felt the car tip and wobble. His driver slowed to a stop while the other cars in the convoy rolled on. The driver stepped out, checked the vehicle, then opened the door to report to his passengers that the car had sustained several punctures, probably made by horseshoe nails in the dust. There was nothing to be done: the group would have to turn back.

As the car rattled along the road, Lekhnovich spied another black sedan traveling toward them at speed, kicking up clouds of dust. As the vehicle approached, he counted four men inside. The man in the passenger seat signaled to Lekhnovich's driver to stop. The cars pulled alongside each other.

"Where is Academician Vavilov?" the stranger demanded through an open window.

Lekhnovich answered straightforwardly. He laid out the expedition's planned route, then asked why they wanted to know.

"He has taken with him from Moscow some important documents concerning the export of grain that are urgently needed," the man explained. Then the mysterious car roared off. The encounter did not strike Lekhnovich as unusual or parlous. If anything, it showed the Plant Institute's work was considered sufficiently important for Moscow to dispatch delegates to collect Vavilov's documents—an encouraging sign. Back at the hostel, Lekhnovich retreated to his room and waited for Vavilov to return from the mountain, not sensing the stirring of a betrayal.

III

VAVILOV'S NATURAL FERVOR FOR PLANT hunting, which kept him digging through the long, hot hours, was not merely theoretical. It was rooted in a desire to reap tangible, hunger-defeating benefits from his work, partly inspired by the close memory of generational poverty in his own family. His grandfather had been one in a lengthy line of *muzhiki*—Russian peasants—but Vavilov's father had worked his way from errand boy to manager of a merchant's store to become a factory owner. In this way Vavilov's father, who was bright but self-taught, raised the family out of serfdom into a position of relative comfort and wealth that facilitated his children's educational prospects and instilled in each of them a scrambling drive for success.

The Vavilov children were each drawn to science. One of Vavilov's sisters became a doctor, the other a bacteriologist, and his younger brother a physicist. Vavilov's inquisitiveness about the natural world drew him to biology. In 1906 he joined the Timiryazev Academy of Agriculture in Moscow.[*] He developed the capacity for arduous work at the academy, where lectures began on September 15 and continued, without break or holiday, until July 15 the following year. This period of intense study was followed by two months of practical work either on farms or experimental stations—a grueling timetable.

---

[*] Also known as the Petrovsky Agricultural Institute, and the Moscow Agricultural Institute.

Even at this early stage Vavilov's interest in plant genetics was rooted in the practical. He drafted his undergraduate thesis on the destructive impact of slugs on food crops in the fields surrounding Moscow. He developed a longing to see his theoretical work produce material benefits. He learned that Russian farmers reaped the poorest harvests anywhere in Europe, less than half the yield of grain per acre than their French equivalents, and less than a third of that harvested by the Germans.

Vavilov knew that around half the harvest depended on the quantity of fertilizer used to feed the crop, and a quarter on the method of cultivation. The final quarter, however, depended on the quality of the seed grain. The young biologist perceived an opportunity: if he could improve the varieties of grain—higher yielding, better adapted, and more resistant—it might be possible for Russian farmers to produce harvests that rivaled those of their foreign counterparts.

IV

IN 1913 VAVILOV VISITED ENGLAND to study alongside the pioneering geneticists Rowland Biffen at Cambridge University and William Bateson at the John Innes Horticultural Institute in London. Vavilov worked out of Charles Darwin's personal library, surrounded by the dust and correspondence of scientific history. Bateson, who had coined the term *genetics* just eight years earlier, had a profound influence on Vavilov's thinking, particularly the idea that potentially valuable wild varieties of wheat, rye, barley, and other crops had been overlooked by farmers in bygone centuries. Bateson believed these snubbed plants might carry invaluable genetic qualities that, according to Gregor Mendel's theory that plants—like people—inherit qualities from parentage rather than circumstance, could be bred into today's crops.

Captivated by Bateson's theory, Vavilov became a botanical treasure hunter. He mounted a series of expeditions to collect and catalog ancient, domesticated varieties of wheat, barley, peas, lentils, and other crops known as landraces. He also sought their wild relatives, which, he reasoned, might prove useful in his experiments to breed supercrops. "In order to guide our breeding work scientifically . . . we must go to the oldest agricultural countries, where the keys to the understanding of evolution are hidden," he said.

In 1916 Vavilov mounted his first major expedition to northern Iran to study cereals. Five years later he visited the United States, then in 1923 he collected a thousand plant samples in Mongolia. The next year he became the first European to lead a caravan across Afghanistan. In 1926 he visited first the Mediterranean, then Italy, followed by the Middle East, and onward to western China, Japan, Taiwan, Korea, Cuba, Yucatán, Peru, Bolivia, Argentina, Uruguay, Trinidad, and Puerto Rico.

Vavilov's face became bronzed with experience. He felt the hot breeze of the Afghan valleys, the buzzing humidity of the Brazilian rainforest, the dry, oppressive heat of the Ethiopian plains. Despite his growing reputation, Vavilov took a humble, openhearted approach on his travels. When he arrived in a new region, he would attempt to learn the local language and approach peasants not as his inferiors, but as colleagues from whom he could learn why particular seeds had been chosen as suitable for their regions and climes.

Vavilov was motivated by scientific curiosity, not profiteering. In this spirit, the specimens he collected were freely given and respectfully taken, or purchased with expedition funds. He made the effort to learn local dialects; Iranian land-owners known as *deqhans* were delighted when he talked to them about early farming practices in impeccable Persian. He would send seeds, spikes, and the stones of exotic fruits by post to Leningrad. While touring the Amazonian riverbanks he collected butterflies with monstrous and intricately patterned wings, which he carefully packaged up and sent home alongside packets of seeds. And even as he passed into middle age, he approached problems "in a spirit of youthful inquiry and optimism." He always included his staff in his adventures, sending postcards to the seed bank describing the plant varieties he had seen.

Vavilov's pioneering work, combined with his eagerness to collaborate with international researchers (he often spoke of the Institute's need to be "on the globe," a term he coined to emphasize the benefits of a strong international reputation), eventually garnered him a slew of prestigious awards. In Britain he was an elected member of the Royal Society of London, the Royal Society of Edinburgh, and an honorary member of the Linnean Society of London and the Royal Horticultural Society, and of the Royal Society of Biology. In the United States he became a member of the American Geographical Society, and an honorary member of the Botanical Society of America. He was awarded honorable associations and honorary doctorates in Germany, India, Czechoslovakia, and Bulgaria. In 1932 the British geneticist and Vavilov's colleague at Cambridge Rowland Biffen declared that under Vavilov's direction the Soviet Union had become the world leader in plant breeding, providing an instructive example of how countries might protect their populations from famine and starvation.

By 1934 Vavilov had established more than four hundred research institutes and numerous stations around the Soviet Union. His journal, the *Bulletin of Applied Botany, Genetics and Plant Breeding*, had become a leading international publication in its field. Where other Soviet scientists subscribed to journals, Vavilov often obtained papers firsthand from the authors whom he called friends. A charismatic, sought-after lecturer, he had indisputably become the most famous and influential biologist-explorer in the world. A rangy, tireless seeker of knowledge and truth, his accomplishments embodied the essence of the scientific spirit.

V

NOT EVERYONE WAS ENAMORED OF Vavilov and his theories, however. One of his former pupils, a peasant horticulturalist named Trofim Denisovich Lysenko, followed Jean-Baptiste Lamarck's early-nineteenth-century theory that organisms could acquire traits in their lifetimes from their environments. These qualities would then be passed down to the next generation. There was no need for genetic engineering or seed banks, which, Lysenko argued, represented a waste of time and resources: one simply had to train plants to meet one's goals, a theory he named vernalization.

Lysenko's outlier theory resonated with the country's leader, Joseph Stalin, who approved of the idea that plants, like workers, could be transformed by an act of political will. Stalin also liked that, unlike Vavilov, Lysenko came from peasant stock, and that his theories did not rely on academic laboratory work. Better still, at a time when millions of Russians and Ukrainians were perishing from starvation caused, in part, by poor crop yields, Lysenko promised Stalin that he could meet the demand for improved crop varieties within three years, seven fewer than Vavilov estimated his work required to produce results.

Stalin was under pressure to undo the effects of the three-year-long terror famine his policies induced in 1932. So when, at a 1935 conference, Lysenko delivered a speech in which he vilified the scientific elite and promised quick-fix solutions to the problems of Soviet food production and distribution, his message resounded. Before the speech was even finished, the Soviet leader rose from his seat and shouted, "Bravo, Comrade Lysenko, bravo!"[*]

Vavilov followed Lysenko's work closely but suspected his emerging rival had manipulated the results of his experiments to support his ideas. Vavilov's students ridiculed Lysenko's apparent ability to "produce a camel from a cotton seed, and a baobab tree from a hen's egg." Nonetheless, propelled by the winds of Stalin's favor, Lysenko sailed past his former teacher through the ranks of the Soviet hierarchy.

Vavilov had begun to experience powerful opposition in the late twenties as part of the so-called Cultural Revolution, Stalin's attacks on the intellectual elite designed to suppress those who might threaten his regime with dissenting views. Lysenko's arrival on the political-historical stage hastened the schism. The seed bank was increasingly viewed as a wasteful drain on the state without tangible benefit. "Practical farming has not been made easier because this [seed collection] is stored in

---

[*] At another conference where both Lysenko and Vavilov were due to speak, Stalin praised Lysenko's presentation, then left the room when Vavilov stood to speak.

the institute's cupboards," said Vyacheslav Molotov, the formal head of the government. Others dismissed Vavilov's expeditions as little more than expensive luxury tourist trips. "In order to assemble 300,000 specimens for his collection of world plant resources, the Institute of Plant Breeding had to organize dozens of excursions to every part of the Old and New Worlds and spend millions on them," wrote one critic at the time. "And what benefit did plant breeding derive from that? Not a thing."

It was Vavilov, however, whose reputation prevailed internationally. His expeditions were covered by Western journalists, and, on his travels, he befriended dignitaries and world leaders. It was Vavilov whose achievements the *New York Times* celebrated in its pages. One English biologist wrote of his Russian counterpart, "His unsleeping mind, his untiring body, his ambitious plans, even his flamboyant showmanship, are all Napoleonic in character." In Stalinist Russia, to be acclaimed by so many international writers and intellectuals could soon become a problem.

Vavilov suspected that his close ties to Western science had brought him under the surveillance of the Soviet security services. In letters written to his American friend Harry Harlan, he began to use coded language to signal danger. All around him science seemed ever more deeply politicized. He faced criticism for hiring staff to work at the seed bank regardless of their social background and Party affiliation. His mentor, William Bateson, was dismissed by Lysenko's collaborator Isaak Prezent as a "fascist" and "racist"—scandalous accusations designed to tarnish Vavilov's reputation by association.

The collection of seeds, too, soon came under attack. In October 1937 *Pravda* published an editorial that claimed "[Vavilov's] expeditions have absorbed huge amounts of people's money. We must declare that practical value of the collection did not justify the expenses." One of his close friends, Dr. Anaida Atabekova, was fired from her post at the Timiryazev Academy for conducting "seditious" research—namely, the study of the effect of X-rays on plant matter. As Stalin began to imprison intellectuals on charges of being "enemies of the state," banishing them to labor camps to be "reeducated" in accordance with Communist principles, Vavilov wondered for how long he could lead the Plant Institute in such an oppressive climate.

Dedication to his research made Vavilov incautious. He disregarded the political sensitivities of his international collaborations. Even after the Molotov-Ribbentrop Pact of August 1939, an agreement that made Britain an enemy of the Soviet Union, Vavilov continued to collaborate with British geneticists. A few months before he traveled to Ukraine, he approved a plan by the biologist Cyril Darlington to translate the seed bank's latest volume on genetics into English. The flurry of correspondence between Leningrad and London swelled Vavilov's file kept by the People's Commissariat for Internal Affairs, the NKVD.

Vavilov knew his actions were provocative. In a sense of heightened paranoia, he began to call home every time he departed one place and arrived at a new one. But his conviction that discipline, not politics, should inform research and scientific collaboration did not waver. At a March 1939 staff meeting he delivered a speech that signaled his willingness to cleave to the first principles of scientific inquiry, no matter the cost : "We shall go to the pyre," he said. "We shall burn. But we shall not retreat from our convictions."

Later that year, while Vavilov was away on a scientific expedition, Lysenko replaced the twenty-seven senior scientists who served on the Plant Institute's scientific council with his own followers. When Vavilov returned, he offered his resignation, saying it was "impossible for me to continue as leader of the Institute." The request was rejected. If Vavilov was to lose his job at the seed bank, it would not be his decision to make.

VI

IT WAS DARK WHEN VADIM Lekhnovich and his colleague Professor Fatikh Bakhteyev returned to the hostel after taking dinner in town. The men were unconcerned; their leader would return late from a seed-collecting excursion. Lekhnovich approached the doorman on the gate and asked whether he had seen Vavilov. Yes, he replied, the botanist had returned a little while earlier. But as he was unloading the car, another vehicle containing a group of men had pulled up and demanded he accompany them for an urgent phone call with Moscow. Vavilov departed in such a hurry, the doorman said, that he had left his rucksack on the pavement.

Lekhnovich and Bakhteyev took the bag from the doorman and returned to their room to await Vavilov's return. The botanists emptied the sack of samples their leader had collected that day and checked each specimen for damage. Among the plants they removed was a sample of grain they did not recognize. Could this be one of the original wheat plants brought to Europe from Egypt four or five thousand years ago?

At around midnight there was a rap on the door. Two young men entered the room. One gave Lekhnovich a note written and signed in Vavilov's unmistakable hand.

In view of my sudden recall to Moscow, hand over all my things to the bearer of this note. 6. 8. 40. 23:15 hours. N. Vavilov.

Vavilov was already at the airport, the stranger explained, awaiting a plane to Moscow. The botanists hurriedly gathered his belongings, discussing what he might need for the unexpected trip, and what they should keep and return to Leningrad. Growing impatient, the strangers insisted that Vavilov needed everything; nothing was to be left behind.

With the bags packed, the botanists carried Vavilov's luggage outside to a waiting vehicle. A third man occupied the driver's seat. Lekhnovich realized there would not be room for both him and Bakhteyev to accompany the cases to the airport. Lekhnovich volunteered to stay behind. His colleague walked around to the side of the car and moved to open the door. "Is it worth your while going?" asked the stranger from Moscow sharply.

"You must be joking?" Bakhteyev replied. At least one member of the expedition needed to meet with Vavilov before he departed. How else would they know what they were supposed to do for the remainder of the trip? As Bakhteyev placed his hand on the door handle, the stranger landed a sharp blow to his head. The botanist fell backward away from the car and landed heavily on the ground. Before he could recover, the car doors slammed in quick succession, and the vehicle carrying Vavilov's bags and officers Kobstev and Koslov, agents of the NKVD, accelerated away into the night.

Red Army soldiers lift crates containing art belonging to the Hermitage onto trucks to be loaded onto the trains that will evacuate the valuables from the city. Thanks to the preparedness of the museum's director, Professor Iosif Orbeli, the first artifacts departed the city just nine days after the invasion began. Staff at the Plant Institute did not feel empowered to follow the museum's example.

II

UNWANTED TREASURE

I

ON THE MIDSUMMER EVENING OF June 21, 1941, passersby gawped at the meticulous displays in the shop windows of Leningrad's glorious boulevard Nevsky Prospekt. During the summer solstice, residents of the northern metropolis enjoyed almost twenty-four hours of sunlight. Professional footballers in the city drew their curtains and took to their beds early in anticipation of their Sunday fixture at the Dynamo Stadium. Dancers rested their calloused feet ahead of the State Ballet School's afternoon performance at the Mariinsky Theatre. Many of the city's apartments sat empty, their occupants away at their dacha holiday homes in the leafed and pleasant countryside.

Dr. Nikolai Rodionovich Ivanov, the Plant Institute's resident bean expert and one of Nikolai Vavilov's closest and most faithful associates, joined his wife, Klavdiya, in bed, excited about tomorrow's plans. Ivanov, who was tall with thick eyebrows and intense eyes, intended to spend the following morning in his study working on his dissertation. Then, with his work complete, he could guiltlessly travel to Pavlovsk, a picturesque rural town south of the city, to enjoy an afternoon walking with Klavdiya in the sun among the orchards whose trees had been grown from seeds collected by Vavilov. Ten months since his fateful trip to Ukraine, the Institute's director remained missing, his whereabouts still unknown. But his colleagues were surrounded by reminders of his work and the mission he had carefully instilled in them.

A thousand miles south of the sleeping city, three million German soldiers were stationed along an eleven-hundred-mile front that stretched from the Baltic Sea to the Balkans. The massed army stood poised to invade Russia. While Ivanov slept in Leningrad, a German soldier stole away from his unit and plunged into the Bug River near Sokal, Ukraine. As his compatriots enjoyed a whispered prebattle prayer in the twilight, the soldier, a Communist sympathizer, swam to the far bank, then climbed into Soviet territory in his sopping uniform and frantically searched for a Russian he might warn of what was about to happen.

In fact, Russia's leaders had already received dozens of warnings that their country's alliance with Nazi Germany was about to fail. Russian agents posted in Europe had repeatedly reported that an invasion was imminent; German reconnaissance planes had been seen violating Soviet airspace, and panzer tanks had been spied rolling up to the border. This information had been rejected by the Kremlin. A few weeks earlier a Soviet spy based in Tokyo sent a microfilm transcript of a conversation between the German foreign minister and the German ambassador to Japan that explicitly set the time of the invasion to mid-June. Stalin's response was to denounce his agent as "a little shit." Such reprisals stifled the transfer of useful information. When the Soviet ambassador in Berlin was handed a German army Russian phrase book containing the terms *Surrender!*, *Hands up!*, and *I'll shoot*, he decided to keep it to himself.

When, a few hours later, news of the German deserter's warning reached Moscow, Stalin insisted to his senior generals that conflict could still be avoided. The Red Army could be brought to a state of combat readiness, but no more. Besides, he pointed out, this deserter might have been sent to provoke a conflict. After midnight, the Russian leader clambered into an armored-plated limousine and, together with his entourage, headed to his dacha at Kuntsevo, where the group drank, ate, and watched films together in an atmosphere of feigned ignorance.

II

IVANOV AWOKE THE NEXT DAY to a warm, welcoming morning full of great promise. He ate breakfast, then took to his study to begin the work he hoped to clear before he could relax. The telephone rang. After a brief conversation Ivanov gravely replaced the receiver, his mind a chaos of competing priorities. Klavdiya asked what was wrong. Ivanov could reply only, "War."

At 0400 that morning, just as the deserter had warned, a hail of artillery shells began to fall on Russian border defense targets, bridges, command posts, and power stations. German bombers had attacked the Soviet cities of Kiev, Minsk, Vilnius, and Sevastopol, while the amassed army swarmed across the border, accompanied by more than three thousand tanks and half a million horses.

To most Russians the preceding twenty-two months of the Second World War—the conflict that would now become known in Russia as the Great Patriotic War—had felt like a distant skirmish, something happening somewhere far over the western horizon. While Nazi tanks rolled across Europe and into France, life in and around Leningrad had continued unaffected. Many assumed the nonaggression pact Stalin had signed with the German leader Adolf Hitler in 1939 would hold until the war ended.

Now, as the city's residents gathered around its network of loud-speakers to hear the minister of foreign affairs announce news of the invasion, listeners perceived the note of surprise in his voice. Despite weeks of warnings, the government had seemingly been taken un-awares. Stalin's man in Leningrad, Andrei Zhdanov, was absent from the city, having just commenced—at Stalin's suggestion—a six-week holiday at Sochi, the Black Sea resort. Most Leningraders had assumed that momentum would carry this war only westward. Thus, everyone from the local grocer to the nation's leaders was profoundly unpre-pared.

Ivanov set out from the old merchant house he and his wife shared on Vasilyevsky Island, one of the oldest parts of the city, and crossed the Lieutenant Schmidt Bridge toward the seed bank to which the thirty-nine-year-old botanist had dedicated his career. By now Ger-man troops had overwhelmed the meager resistance they met on the border to plunge deep into Soviet territory. But Leningrad appeared unchanged. Shop shelves sat well stocked. Clear water flowed from willing taps. Trams ran to dependable schedules, ready to carry commuters to their homes, where stoves bubbled with stew. The rhythms of familiar routine would soon stall, to segue into a new song of as-yet-unknowable agonies. But like anticipatory grief for a dying loved one, misery was for now a shapeless, looming thing; something one hoped might still be avoided.

Ivanov crossed St. Isaac's Square where, before the Revolution put an end to organized religion, worshippers had once gathered inside the gold-dipped cathedral. Then he strode through the tall wooden doors at the entrance to 44 Herzen Street. As he climbed the majestic staircase that fed a rabbit warren of shadowy corridors, the botanist felt reassured. Inside the Institute time appeared to flow differently. This was a place of profound stillness. Something of the essence of life had been captured and stored in these rooms, genetic material that had outlasted generations of shape-shifting human conflict, cycles of politicians, successions of tsars, recurring battles for territory and resources. There was a reassuring continuity to the collection, a connection to the genetic fabric of the natural world, which predated the arrival of humankind, and which might yet outlast the species' violent squabbles and vanities.

For all the magnificence and history held within these walls, the Institute's rooms were poky. A parquet maze of flooring met walls covered in thick layers of accumulated paint, and tired wooden paneling. There was a sense of managed dilapidation. Despite its scientific importance, this was not an Institute awash in resources. Four years earlier, after the seed bank suffered a reduction in funding, the Institute's publishing house had closed. Still, here in the heart of the enterprise, the air remained dense with the feeling of study. Desks lurked beneath a burden of scientific clutter: sheets of paper scrawled with notes, stuffed envelopes, tilting lamps, hunchings of microscopes. Inside the rooms that held the precious seeds, shelving units clad the walls and bisected the floor. These were laden with identical-size metal containers—around 120,000 in total—each labeled with a string of numbers used to identify the contents.

Ivanov believed in his vanished friend and leader's mission: to preserve the history of the planet as encoded in the DNA of its plant life. He viewed the seeds stored within the Institute's walls as more than just packets of potential growth. They were living artifacts of a time and place, which carried within them the secrets of their origins and the conditions of their growth. The Plant Institute stood as a reminder that the past is not dead, but persists in living things, waiting to be unlocked and explored by those with the curiosity and will to do so. The building was a Noah's ark of plant matter, a catalog of the past and a hope for the future, an insurance policy against humanity's indifference and negligence.

The tins sat like miniature bunkers. They contained treasures that had been carried hundreds, sometimes thousands, of miles to Leningrad. There was naked-grained barley found on the plateau that borders Turkestan, India, and Afghanistan; wild perennial flax picked from Iran; orange and lemon pips collected on the road to Kabul; radishes, burdock, edible lilies, and chrysanthemums from Tokyo; sweet potatoes from Taiwan. There were Korean soybeans, Spanish gorse, Egyptian clover tobacco. Each specimen had been handpicked by Vavilov and his team, wrapped in tissue paper, then brought to Leningrad in burlap sacks or roomy pockets.

Some seeds were small and furred, others fat and smooth. To the trained eye these kernels held all the wonder of an artist's masterwork. Still, no sightseer stopped by to gasp at Vavilov's accumulated wonders. Unlike the expansive canvases and handsome pots held in the city's art galleries and museums, the Institute's riches were on a meeker scale: latent life encased in tiny shells, then rested on dry tissue paper, and stowed in anonymous tins.

Ivanov could not imagine the Nazi invaders would ever make it as far as the outskirts of Leningrad, let alone here to the heart of the city. And even if an army were to storm these rooms, what soldier would understand the value of the little packets? Still, like thousands of others, Ivanov decided that he would enlist and help ensure Russia's enemies would never make it as far as the city, and inside these walls.

III

ACROSS THE SQUARE THE ART historian Professor Iosif Orbeli closed his office door and clambered up the staircase that led to the long corridor that flanked the galleries of the Hermitage Museum. The museum's director stopped by a window to survey the city, with its stuccoed white buildings, soaring spires, arced bridges, and glittering palaces. With his great white beard, abundant brows, and incandescent eyes, the fifty-four-year-old had the appearance of an Old Testament prophet, a fitting description today, when he felt as though he had foreseen a tragedy to which everyone else was oblivious.

The Hermitage had opened as usual at eleven o'clock, its workers still unaware of the German invasion. The museum guides had gathered their tourists and led them from room to room, gesturing at the Renoirs, the da Vincis, and the Raphaels, a density of artistic splendor that could inspire in any visitor feelings of both transcendence and fatigue. An hour later Orbeli and the museum's guards, researchers, and other assorted staff members had gathered around the radio and listened to the news of the invasion.

In his office Orbeli had first placed a call with the Committee on Arts in Moscow: What, he wanted to know, was the plan to save Russia's greatest collection of artworks from destruction or, should one day Leningrad fall, looting? But during the six calls he made during the next two hours, nobody in Moscow could provide clarity or instruction. If plans had been made for such an invasion, it seemed those responsible for protecting Leningrad's treasures, be they botanical or artistic, did not feel sufficiently empowered to initiate them yet.

At the window, as he surveyed the city skyline, Orbeli caught sight of a gray sausage-like antiaircraft balloon rising beyond the spire of the Peter and Paul Fortress. At least someone in the city seemed willing to act. The director made a decision. He would not wait for Moscow's leaders to organize themselves into whatever *matryoshka*-like configuration of superiority was necessary to relinquish clear guidance. The evacuation of the museum would begin at once.

Unlike Stalin, Orbeli had planned for this day. Four years earlier he had deployed a team of joiners to construct wooden boxes of varying sizes that could be used to transport the most valuable exhibits if the museum became threatened. The team produced thousands of crates, secretly housed in the Sampson Cathedral, a Russian Orthodox church situated across the Neva River. Museum workers had also stockpiled three tons of cotton wadding, ten miles of oilcloth, and fifty metric tonnes of wood shavings—supplies that could be used to cushion artifacts if they needed to be hastily packaged for safekeeping. The precise logistics for how two and a half million rare objects—including fifteen thousand paintings, twelve thousand sculptures, sixty thousand drawings, half a million archaeological monuments, and more than a million coins and medals— would depart Leningrad could be figured out later. For now, Orbeli's task was merely to prepare the museum's vast collection for transit.

Now the moment had come to act. The curator ordered the guards to close the museum and refuse further public admissions. He ordered every employee to begin removing the most precious paintings from the walls, ready to be placed in the steel-lined vaults that housed the museum's collection of Scythian gold. Researchers, security personnel, and technical employees all helped with the packing. There was none of the awe that a person might in other circumstances experience when handling masterworks at such proximity. The focus was on speed and safeguarding.

The packers entombed valuables in tar paper, cellophane, oilcloth, cardboard, and wrapping paper to deter damp and dust. Two women fed pieces of crumbled cork through a crack in the lip of a fourth-century Chertomlyk vase using a teaspoon. Other employees prepared inventory lists and a packing order. When the lid of each crate had been fastened and labeled, workers added it to an accumulating stack on the mosaic floor of the temple-like Hall of Twenty Columns.

Orbeli's staff worked until, as one volunteer recorded, their legs began to throb. In rooms that smelled of floor polish and history, these were some of the first of Leningrad's citizens to feel the stress of war, not as an abstract threat, but as an aching in their muscles. As they wandered from empty room to empty room, they could also feel war as an absence. "The empty Hermitage is like a house after a funeral," noted one observer.

IV

WHILE THE TEAM AT THE Hermitage forged ahead with their self-made plans for evacuation, Ivanov and his colleagues at the seed bank were unsure of what to do. Vavilov was, like Orbeli, a leader of clear, decisive character. But nobody had heard from him for almost a year. Knowing what had happened to his predecessor, Johan Eichfeld—who had assumed directorship of the Institute in Vavilov's absence—did not want to act without guidance. Leaderless and lacking clear instruction from Moscow, the seed bank staff felt lost.

Vavilov's disappearance from the Plant Institute represented, for his supporters, a loss of still unknown shape and impact. News of his disappearance had reached the building some ten months earlier, on August 12, 1940, when Bakhteyev returned to Leningrad. Confusion reigned. That month the organizers of the All-Union Agricultural Exhibition awarded Vavilov a gold medal for services to Soviet agriculture. Why, then, would the authorities simultaneously arrest him? Some staff believed that Vavilov would return at any moment and once again stride through the laboratory doors in one of his smart suits to resume his fervent work. But within a few weeks Vavilov's loyal supporters had become so convinced of his arrest under false charges that they drew up letters to the Central Committee of the Communist Party, the government, and the NKVD, vouching for his character and declaring that he was no spy. Vavilov should, they urged, be released at the earliest opportunity and allowed to continue his important scientific work. Nine Institute staff members signed the letter. But when Nikolai Kovalev, one of the signatories, showed a draft to a relative who worked for the NKVD, he received a warning that anyone who put his or her signature to the letter would be arrested for supporting a suspected "enemy of the people."

The group discussed their options. As the only member without young children, the biologist Nina Basilevskaya volunteered to bear the full weight of the threat; hers would be the only name on the letter. A summons to the Communist Party headquarters in Leningrad promptly arrived. There, she pleaded Vavilov's case, stating that if the seed bank's director was banished to a gulag, then most of his staff would follow him into exile. It was a risky wager intended to demonstrate the depth of affection Vavilov's staff harbored for their leader.

"Nonsense," a Party official replied. "We do not make mistakes when it comes to arresting people. Get on with your work quietly, and don't bother officials with such trifling matters."

Three days later Basilevskaya felt the dull, irresistible force of official reprisal when she was dismissed from the Institute. Kovalev's arrest soon followed. Undeterred, Vavilov's supporters prepared a new document, several dozen pages long and typed in secret, which laid out the material benefits the Soviet Union had enjoyed from Vavilov's expeditions, experimental work, and research.

Maria Shebalina, from the department of fodder crops, set out to Vasilyevsky Island with the document in hand to seek an audience with Vavilov's younger brother, Sergei, director of the Optical Institute. If Sergei could pass this trove of evidence to the right person, Shebalina and her colleagues believed, the authorities would release Vavilov in the interests of science.

Shebalina entered Sergei's office and saw Vavilov's younger brother at a large desk. He motioned for her to sit down opposite him. Shebalina noted Sergei's "indistinct, colorless manner of speaking," and his "empty, tired eyes." She slid him the file. Sergei ignored the document.

"There is no misunderstanding," he told her of his brother's arrest. "We can hardly hope to do anything about it."

Naïve hopes for his return further dimmed when, four months later, news reached the seed bank that Vavilov had been dismissed from his position as director of the Plant Institute. Shortly thereafter, police arrived to search his office, then his apartment on Nevsky Prospekt. On hands and knees plainclothes agents removed the Institute's floorboards and raided the seed bank's cellars and attics. When a bewildered staff member inquired as to what they were searching for, the agent replied, "Bombs."

A bogus story about the circumstances of Vavilov's arrest began to circulate among the city's scientific community. The botanist, the story claimed, had visited Ukraine with a plan to cross the border and flee to the West, taking his scientific knowledge and findings with him. It was a nonsensical tale. Vavilov had enjoyed hundreds of opportunities to emigrate during the previous decade. He had always returned to the building in the heart of Leningrad that housed his family and life's work.

So began a systematic attack on the seed bank as the authorities promoted scientists loyal to the peasant farmer Trofim Lysenko, such as Eichfeld, to senior positions in the Institute and dismissed Vavilov's highest-ranking supporters. There were a few ardent anti-Vavilovists among the researchers and doctorate students at the seed bank, but this minority felt emboldened by the state's persecution of the Institute's erstwhile leader. Conferences held at the seed bank became fraught. After one especially impassioned meeting, the head of the Institute's biochemical department, an elderly professor named N. N. Ivanov, returned home, lay down on his bed, and promptly died.

Now, with the German armies tearing toward Leningrad at unknown pace, none of Vavilov's loyal staff members felt sufficiently empowered to copy the Hermitage's lead and pack their vanished leader's collection for evacuation.

V

ON THE FIRST NIGHT FOLLOWING the invasion, Leningrad's air-raid sirens blared, joined in braying, hellish chorus by factory hooters and the foghorns of ships docked in the Admiralty Shipyard. The city lay more than seven hundred miles from Brest-Litovsk in Belarus, where the first German soldiers had crossed into Soviet territory. But many residents instinctively felt that, as the former capital and center of the country's culture and industry, their city would rank high among the Nazi leadership's strategic objectives.

Operation Barbarossa,[*] the German plan for the invasion of Russia, had been ratified by Hitler six months earlier, on December 18, 1940. Having occupied Poland and France, Germany's leader intended to annex Russia from the border with Ukraine to the east of Moscow, around one-eighth of Russia's total territory. It was to be a three-fronted war, waged across an unimaginably vast theater: the largest country on earth, representing a sixth of the planet's landmass, home to ten time zones at the time. Army Group South would cut toward Ukraine, territory rich in industrial, mineral, and agricultural resources; Army Group Center would roll eastward toward the capital, Moscow; Army Group North, led by the brilliant military strategist Field Marshal Ritter von Leeb, would take aim at Leningrad, the second-largest city in the Soviet Union and a site of both material and symbolic significance. Hitler optimistically believed the takeover would be complete within four months.

The Nazis planned to use the Soviet Union as a vast repository of resources, plundering it to fuel the German war machine and feed its civilians at home. Before the invasion, Russia had diligently sent grain to Germany, diverting valuable resources toward a nation that would soon become its enemy. Now that the countries were at war, this kindness would not be reciprocated. German soldiers had been instructed to requisition food from local populations even if it meant locals would starve. Herbert Backe, state secretary in the Ministry for Food and Agriculture and the man at the center of Nazi food and distribution policies, urged the German soldiers to harden their hearts toward Russian civilians and show neither "false sympathy" nor "misplaced pity." The Russian, he wrote in a June 1941 memo, "has already endured poverty, hunger and frugality for centuries. . . . The Russian stomach is stretchable."

Hitler ordered that no unit was to attack Moscow before Leningrad fell. Situated on the Baltic Sea, a hundred miles southeast of the border with Finland, the city was seen by Hitler as the strategic key

---

[*] Originally code-named Operation Fritz; Barbarossa was named after the twelfth-century German emperor who launched a crusade against the Slavic people.

that would unlock the door to Eastern domination. Leningrad's six-hundred-odd industrial factories rivaled Moscow's in number, and the city was the home base of Russia's Baltic fleet. Leningrad would link Germany's forces with those of Finland and sever the supply lines that carried Allied aid from the arctic ports of Archangel and Murmansk.

Once taken, Leningrad would provide Nazi Germany with agricultural land and oil wells, access to the Soviet Union's key shipyards, steelworks, and major arms manufacturers. Hitler's designs on the city were not merely utilitarian, however. As the birthplace of Revolutionary Communism, a hotbed of liberal intelligentsia[*] and incubator of perversion, Leningrad symbolized all that Nazi ideology opposed.

To take Leningrad would eliminate a major center of intellectual and artistic endeavor and facilitate the capture of the many priceless artifacts held in its various institutions. In Hitler's delusional, surreal, and impractical view, the city was a prize worth almost any cost. As he put it in an order read out to troops an hour before the invasion began, "You are going into action in order to save the whole of European civilization and culture."

VI

THE NEXT DAY CITIZENS BEGAN to board up public monuments or bury them in the ground. The city braced itself for impact. Leningrad's administration had neither the autonomy nor the systems required to organize the hundred thousand volunteers who turned up to pledge their support, Ivanov among them. Recruitment centers were mobbed; lines of patriotic hopefuls trailed through double doors onto the streets. It took hours to reach the front desk; many volunteers filled out their forms while they waited in line, resting their paper on stair-

---

[*] In his sparkling memoir, *Speak, Memory*, the St. Petersburg–born novelist Vladimir Nabokov argues that, by the time of the invasion, "all liberal-minded creative forces—poets, novelists, critics, historians, philosophers and so on—had left Lenin's and Stalin's Russia." The exceptions were, in his estimation, "either withering away there or adulterating their gifts by complying with the political demands of the state." If so, the siege of Leningrad would have a revitalizing effect on some of these "withered" creatives, as the clear-eyed reportage of Lydia Ginzburg and the resonant poetry of Olga Bergholz attest.

cases, or windowsills. When Ivanov reached the desk, he was told to return to the seed bank and continue his work as normal; he would be called upon if required.

Many citizens chose to believe the government's soothing assurances that the Red Army was ably holding back the Germans, that the city's food stocks were plentiful, and that its advanced antiaircraft systems could see off any German pilot who dared fly into range. There were signs that this might not be the entire story, however.

From the first day of war Leningraders could withdraw no more than two hundred rubles from their savings, a maddening imposition as the prices of essentials began to rise precipitously. A few days after the invasion, the government mandated nightly blackouts, leading women to drape blankets over window openings to hoard heat and light. The authorities issued children rose-shaped phosphorous badges to help their parents locate them in the dark. Police confiscated radios to ensure the population was not at risk of being subjected to enemy propaganda.

Still, these impositions could mostly be dismissed as the inconveniences of war. The German army was hundreds of miles away and comfortably outnumbered by the Red Army's troops that stood in their way. Nobody had any inkling that, within three months, Leningrad would become the setting for the longest siege in recorded history.

An air-defense soldier watches for incoming German air attacks in 1942. In the early months of the war, musicians from the Leningrad Conservatory, believed to have superior hearing, would stand among the columns at St. Isaac's Cathedral, close to the seed bank, and listen for aircraft engines, calling on-duty air defense officers with information about the direction of their approach.

III

SPIES AND TRAITORS

I

As the milk light of dawn rose on June 26, 1941, four days after Germany invaded Russia, sentries at a Soviet guard post on the outskirts of the Latvian city of Daugavpils watched a small convoy of trucks approach. Each vehicle sat heavy on its tires, obviously burdened with passengers. The city's bridges—the main road bridge, and a rail bridge about a mile to the west—provided vital crossings of the Dvina River for anyone traveling to Leningrad, three hundred miles to the north. It was a key strategic chokepoint. But the Russian guards had no cause for alarm. German units, they had been assured, were weeks of fighting away. As the vehicles drew closer, the Soviets saw that the lead driver wore a Red Army uniform. This must be a routine movement of friendly troops.

Inside the lead truck, the twenty-six-year-old leader of the group of disguised German commandos cautioned his men to remain quiet. *Oberleutnant* Hans-Wolfram Knaak commanded the Eighth Company of the Brandenburg Regiment. The unit was composed of highly able but misfit men who had been trained in subterfuge—Germans who were fluent in Russian, embarked on an infiltration mission. In contrast to the racially selective policies of other Nazi outfits, the unit specifically sought out soldiers who looked like, or who were, Poles or Slavs.

The insignia of the Brandenburg unit—a dagger and a masquerade mask, with narrow eye slits angled to form a malicious sneer—

suggested its modus operandi. Knaak and his men were masters of what conventional officers often referred to disparagingly as "irregular" warfare. When the unit was founded two years previously, the Army General Staff referred to the unit as, simply, "a company of saboteurs."

The lead vehicle drew to a standstill on the Wilkomierz-Utena road. The driver waved and greeted the sentries in flawless Russian. The company was used to remaining calm in the presence of the enemy; they had been some of the first German soldiers to cross Soviet territory even before the invasion began, disguised as Red Army soldiers, on a mission to sow confusion and misdirection.

"Where are the Germans?" asked the sentry, oblivious of the danger he and his compatriots faced.

"Oh, a long way back," the driver replied.

More small talk. Laughter. Then a rolling crunch as the sentries waved the trucks through the checkpoint, into the city, and toward the two bridges that were vital for the Nazi advance on Leningrad. In the back, Knaak's men readied their weapons.

II

SIXTY-FOUR-YEAR-OLD FIELD MARSHAL RITTER VON Leeb—a man whose ambivalence toward the Nazi project could be overlooked by Hitler on account of Leeb's considerable value as a military strategist and leader—had been assigned to lead the march to Leningrad. With more than half a million soldiers under his command, the Sixteenth and Eighteenth Armies were the poor relations of the German forces; with no motorized transport, 80 percent of their troops had to walk into battle, while horses pulled most of their equipment. Leeb, who was bald, lithe, and wore a tightly cropped mustache, also commanded Panzer Group Four, to which Knaak's commando unit was currently attached. Using these fast, powerful front units, Leeb hoped to surround Leningrad before the end of July, a blitzkrieg that would require the massed army to advance into enemy territory at a rate of nearly two hundred miles a week.

It was an ambitious plan but, if anything, the advance was already ahead of schedule. The Soviet troops that Leeb's men had encountered were unprepared both psychologically and materially. The obduracy of belief among senior political leaders that Russia's pact with Germany would endure had left the Red Army unprepared for combat.

Stalin, too, chose to ignore the warning signs—eighty-four of which he received in writing, from sources as reliable as Winston Churchill, who had decryptions from Bletchley Park—which were also downplayed by the unctuous lackeys responsible for relaying intelligence. While Germany moved men and munitions eastward by rail toward the Russian border for weeks in anticipation of invasion, Stalin had banned any preparations for war, which he feared might be considered provocative. On the day of the invasion Russia was so underprepared that many Soviet infantrymen were handed antique rifles from the First World War, in some cases without ammunition, leaving the men pointlessly deployed and vulnerable.

In the first few hours of war, the Luftwaffe destroyed Russian fighter planes and bombers on the tarmacs of their airfields. By the end of the first day, more than twelve hundred planes had been destroyed at sixty-six bases, three-quarters while grounded. Surviving planes were soon outclassed in the air. The German Messerschmitt could fly more than a hundred miles per hour faster than the Soviet I-16. Most Russian planes did not have two-way radios: they could receive but not relay messages, which made it impossible to make plans of attack in the air. A pilot who wished to deliver a reconnaissance report would first have to return to base. Of the fourteen thousand tanks at the Red Army's disposal, all but two thousand were obsolete models. Stalin's senior officers were ex-cavalrymen; their understanding of tank strategy was poor. Within a few weeks of the invasion almost all Soviet tanks had been destroyed.

The speed and ferocity of the German advance was kept from the Russian people, from Leningrad's factory workers—who had no reason to doubt the misleading stories reported in *Leningradskaya Pravda* that claimed German soldiers were deserting in droves—to even the highest echelons of Soviet command. Official communiqués obfuscated facts and downplayed unwelcome news. By the time Knaak and his disguised men entered Daugavpils hoping to prize open a route through which German troops could surge, the chief of the Leningrad Infantry Academy was completely unaware that German troops had already advanced a hundred miles into Soviet territory.

III

A FEW HOURS EARLIER, KNAAK'S unit of thirty men had broken away from Leeb's troops and closed, at high speed, the 120 miles that separated the German tanks from Daugavpils. It was a hot, dirty drive through an empty landscape. The plan was straightforward: assume control of the city's bridges before the Soviets could detonate the charges they had rigged for just such an emergency. It was a Gadarene dash but one that, if successful, would create a funnel point through which the German army could flow.

Past the checkpoint, the trucks weaved through local traffic. The convoy split into two. One group would attempt to take the railway bridge, the other, led by Knaak, the road bridge. The men wore *halb-tarnung*—Russian greatcoats and side caps with the red Soviet star on the front, to obscure their German uniform beneath.

The first group passed five enemy armored cars where a phalanx of Soviet armored vehicles, against which the saboteurs' machine guns were useless, halted progress. The truck pulled back to the main road. There the Germans disembarked and took up positions along the bank, while one of the officers cut through a cable that he assumed had been laid to detonate charges rigged along the bridge. While the Brandenburgers returned fire, a Soviet artillery shell landed on the bridge, setting off part of the demolition charge. The damage would prevent German tanks from passing.

Meanwhile, the second group drove toward the road bridge. The Russian guards on the west side of the bridge were chatting to civilians as the truck approached and did not see the machine guns aimed at them from its windows. As the passengers gunned the men down, the driver accelerated across the bridge. Too late he noticed an antitank gun positioned on the far bank. It fired on the lead vehicle, causing an explosion in which Knaak was mortally wounded.

The surviving soldiers disembarked, crouch-running into the squall of bullets. Knaak's unit was well trained. Every man had undergone an extensive, five-month-long instruction in reconnaissance, swimming, hand-to-hand combat, demolitions, and marksmanship and was competent in the use of both German and Soviet weapons. The men had been trained, however, in covert infiltration, not shock-troop tactics. The unit dug in, holding its position and preventing several attempts by Russian soldiers to recapture and detonate the rigged charges. Twenty minutes later the first German panzers rolled into the city. At 0800, Leeb received the signal: "Surprise of Dvinsk and Dvina bridges successful. The road bridges are intact."

Knaak was dead, but the saboteurs had achieved their goal. After intense fighting, and the arrival of the Eighth Panzer Division's motorized infantry, the German tanks crossed the river at Daugavpils, creating a fatal hole in the Soviet lines.

"It is unlikely I will ever again experience anything comparable to that impetuous dash," the tank commander Erich von Manstein recorded. "It was the fulfilment of every tank commander's dream."

The road to Leningrad was clear.

IV

Nikolai Ivanov considered himself a friend of Director Orbeli's, as well as a professional neighbor. Together they had both enjoyed evenings at Vavilov's apartment, where the great botanist would entertain the city's curators and scientists, introducing experts from diverse disciplines to one another over glasses of rice wine. The Plant Institute and the Hermitage Museum had supported each other practically, too. Before the invasion Ivanov had helped to transport hundreds of pieces of glass across the square, a gift to the museum from the seed bank to help replace old or broken windowpanes.

The admiration Ivanov felt toward his industrious contemporaries at the museum was tempered, however, by the obvious disparity in how each institute was viewed by the Soviet state. In less than a week following the invasion, Hermitage staff had boxed half a million objects into a thousand crates. As he crossed the Lieutenant Schmidt Bridge on his thirty-minute commute to the seed bank, Ivanov saw military trucks parked snugly between the Winter Palace's Ionic columns, as soldiers loaded them with crates of artifacts to be transferred to a waiting train. In the seed bank, the shelves sat stacked, their curators devoid of a plan, instruction, or equivalent military support.

While the leaders of other institutions across the city prevaricated, either slow to accept their new reality, or unwilling to act without Moscow's say-so, on July 1, 1941, five days after the Brandenburgers captured Dvina Bridge, the first trainload of the Hermitage's principal treasures left the city. To move so many objects so quickly represented an astonishing feat of coordination and effort. Only children had left the city before the art: fifteen thousand young people, evacuated just two days earlier to summer-camp destinations to the south and west of the city.

The Hermitage train, originally intended for the evacuation of machinery from the Kirov defense works, comprised twenty-two freight wagons, more than twice as many used for the evacuee children. The most valuable items were stowed in an armored carriage. Only one painting—Rembrandt's *Return of the Prodigal Son*—was afforded a crate to itself, and all but four paintings were removed from their frames and either packed flat or carefully rolled. At either end of the procession antiaircraft guns stood on six-wheeled flatbed cars designed to carry especially heavy loads. In the interest of security, the train's engineers were not told their destination prior to departure. Another train ran ahead to clear the tracks. For the masters whose works began to wend their way across Russia under armed guard, on an inconceivably long, fifteen-hundred-mile journey toward the coniferous shelter of the Ural Mountains, it would have been an incomprehensible sight.

Orbeli's foresight soon caught the attention of the directors of other institutions beyond the Plant Institute. Georgi Kniazev, director of the archives of the Academy of Sciences in Leningrad, requested that items from the academy's collection be added to the second train-load of art due to depart the city a few days later. Orbeli dispatched Hermitage workers to the academy, where they removed the mosaic portrait of Peter the Great by Mikhail Lomonosov,[*] two eighteenth-century maps of the city, and a gilded casket containing a handwritten order of Catherine the Great. When the first children fled the city, their bereft, inconsolable mothers had chased after the trucks. For Kniazev, the removal of the academy's treasures inspired in him simi-lar feelings of distress. "I saw them off," he wrote in his diary, "in great anguish . . . the way you see off a loved one: a son, a daughter, a wife."

Next, almost half a million books were removed from the Leningrad Public Library. Moscow authorized the removal of Voltaire's Library, the Pushkin Archives, and various exhibits in the Russian Museum. Members of the Pushkin Drama Theatre, the Mariinsky Opera and Ballet, the Com-posers' Union, the Leningrad Philharmonic Orchestra, as well as around a third of the Zoological Garden's rarest animals, including panthers, polar bears, and a rhinoceros, all departed the city during the early weeks of the summer. The seed bank, however, received no instruction to evacuate.

Unable to join the army as a volunteer, and without the author-ity to execute a similar plan, Ivanov had time to contemplate the seed bank he loved, and the recent injustices to which it had been subjected. There were the scathing attacks from colleagues in Russian science, the disappearance of its charismatic leader, and now, seemingly, the institution's abandonment by Leningrad's authorities. Bureaucratic ineptitude played its role in the failure to evacuate both people and institutes from the threatened city. But as other institutions departed week by week, to Ivanov the situation began to seem less like negli-gence than vindictive omission.

The motivation behind this apparent abandonment remained un-clear. Could it be the same anti-intellectual bias that had affected the arts and sciences across Russia, a specific disinterest in the seed bank proj-ect, or had the Institute merely been forgotten in the mounting chaos of

---

[*] *The Battle of Poltava.*

war? Whatever the reason, as the German army tore toward Leningrad, the Institute and its collection faced imminent, catastrophic danger.

V

LENINGRAD'S PEOPLE WERE IN THE dark. Twice a day Sovinform—the Soviet Information Bureau, established two days after the invasion to be the sole authority to release information about the war—issued catastrophically vague reports. News of major defeats came days or weeks after the events. It was impossible to know the precise location of the front and, it followed, how concerned citizens should be. The truth was that in the space of three weeks, hastened by the capture of the bridges at Daugavpils, Field Marshal Leeb's panzer divisions had forged almost three hundred miles into Soviet territory, through Lithuania, Latvia, Estonia, and now into the Leningrad province.

Leningraders knew to not trust the official reports, which were euphemistic and often illogically optimistic. *Nashi byut,* the joke went, *a nemtsy berut*—"We're winning but the Germans are advancing." Scraps of dependable information came from those men who returned from the front, but few were willing to put their faith in hearsay. Ordinary Russians were sufficiently ignorant of the speed of this blitzkrieg that many holidaying children became trapped in towns and villages south of the city, cut off by the German advance.

To anyone who was paying attention, however, it was clear that those in power were preparing for the worst. Throughout the summer, workmen obscured street signs with white paint and removed identification boards from outside prominent buildings, measures intended to confuse saboteurs in the city's labyrinth of canals and roads if they entered the city. Tram conductors no longer announced stops. The city's Communist Party headquarters was draped with camouflaged netting and surrounded by a constellation of machine-gun nests.

Fearing German bombers could use the gilded dome of St. Isaac's Cathedral, visible for miles in every direction, as a landmark to direct artillery strikes, four climbers concealed its gilt surface under a grayish fireproof coating. The Admiralty's dazzling spire received similar treatment, painted by a music teacher and amateur alpine mountaineer named Olga Fersova. Carlo Rastrelli's statue of Peter the Great vanished from its home in front of the Engineers' Castle, leaving only the empty plinth. The Klodt stallions galloped from their position on Anichkov Bridge and now lay buried in the Summer Gardens under mounds of protective earth. Citizens filled their attics with sand and bandaged their windows with gauze or crosses of sticky tape to reduce the risk of shards becoming shrapnel.

To Lydia Ginzburg, the literary critic and faithful Leningrad resident, on a sunny day, and viewed from a distance, the stickers appeared like "cheerful" ornaments. When she looked more closely, however, at the sun-yellowing of the paper, the indelicate blobs of glue, the dirty newsprint showing through, these humble domestic attempts at defense symbolized a kind of death and destruction "which had not yet had time to settle and attach itself."

"The bombing of Leningrad is inevitable," the director of Leningrad's Academy of Sciences recorded in his diary. "Anyone who spreads the opinion that the Germans will not bomb the city is either a dedicated agent provocateur, or a chattering fool."

Soon paranoia gripped everyone. It became illegal to photograph or film on the streets without special permission, and those photojournalists who were allowed to take pictures were fully aware that only those images that showed courage and resolve could be sold and published. Servicemen were discouraged from keeping diaries in case the documents fell into enemy hands. Anyone critical of the authorities could be subject to arrest and, in the most extreme circumstances, sentenced to death by firing squad.

"Spy mania, like an infectious disease, has struck everyone without exception," wrote one resident, who recorded an encounter with an older woman in an oversize mackintosh who grabbed her arm, pointed at a man who wore trousers and a coat of mismatched colors and a suspicious mustache, and shouted, "Did you see? A spy for sure!"

In this atmosphere of powder-keg suspicion, no one was permitted to enter the city without the proper documentation; anyone lacking papers was summarily detained. Soviet propaganda warned the population to be on the lookout for *fifth columnists*, the term given to traitors posing as friendly residents. Clothing considered too Western in style drew misgivings. Misgivings bloomed into denunciations, often and increasingly motivated by the tipster's desire to acquire the target's belongings or even their home. The political right looks for converts; the political left looks for traitors. To some, Communist Leningrad was a city of collaborators.

Swept up in this tidal surge of suspicion, and perhaps grateful for the distraction of focused activity, staff at the Plant Institute established guard posts. Only those directly involved in seed bank work were granted clearance to enter the premises, a measure intended to reduce the risk of spies or thieves. Everyone suspected the Germans wanted to plunder Russia's resources, including its food stocks. Among its vast array of specimens, the seed bank contained strains of grain genetically trained to produce high yields. A soldier might not recognize the value of some old tins and seed packets, but an enemy scientist, the Institute staff concluded, would recognize the latent power in the tens of thousands of tin containers that lined the Institute's walls.

Now, as the German army advanced on Leningrad, Ivanov and his colleagues could do little beside make the most rudimentary preparations. Soldiers had installed machine guns on the Hermitage roof to provide defense if paratroopers landed in Palace Square. The Plant Institute was given no similar defensive capabilities. Instead, the building's manager, Maria Sergeyevna Belyaeva, did her best, placing buckets filled with either water or sand on the landings to douse fires and issuing each colleague a gas mask.

VI

ON JULY 7, 1941, A week after the first Hermitage train departed, the seed bank received its first official directive. Hopes that Moscow had remembered the collection and planned for its removal from the city were immediately dashed. Order 182 merely informed Ivanov and his colleagues that they were to complete their standard daily duties by 1300 each day, at which point they were to report to the Forel Hospital on Stachok Street to help build fortifications until 2200.

These new responsibilities, while necessary for the protection of the city, drew attention and energy away from what the staff saw as their primary obligation: the preservation of the seeds. The sense of institutional abandonment Ivanov and his colleagues felt was compounded by the absence of their leader. Vavilov's disappearance had hit Ivanov especially hard. While Ivanov was in his early twenties he met the famous geneticist at the Leningrad Agricultural Institute in the Faculty of Genetics and Selection. In 1927 Vavilov was appointed to be Ivanov's graduate supervisor. Ivanov saw in Vavilov both an inspirational teacher and a father figure; Vavilov discerned in his new student erudition and gentleness.

Ivanov was offered opportunities to travel—everywhere from Ukraine to the Far East—and scholarly supervision (when Ivanov presented his dissertation on haricot beans to senior staff at the Plant Institute, the academics awarded him a professorial doctorate on the spot). And Vavilov provided his protégé with other forms of mentorship, setting a chair in the corner of his office to allow Ivanov to sit in on meetings, so he might learn diplomacy, as well as botany.

Throughout the recent purges, when Stalin's regime had removed many of Vavilov's most senior supporters, Ivanov had evaded dismissal or redeployment. Now, with the arrival of war, he was promoted to the position of senior research officer and became one of the highest-ranking Vavilov supporters in the beleaguered Institute. After weeks awaiting instruction, finally Ivanov was told to report to Vyritsa, a small town forty miles south of the city, to help dig trenches and antitank ditches.

Ivanov was one of around fifty thousand Leningraders, mostly women and teenagers, who took the train to the southwest of the city to build defenses along the so-called Luga Line. Positioned behind the Luga River, a natural barricade against invaders marching toward the city, this was to be a series of minefields, antitank guns, and barriers, ten miles deep and 185 miles long, with only a small gap between Luga and Gatchina, through which the Red Army could retreat if necessary.

Volunteers were needed everywhere, as evidenced by the dozens of competing notices posted on factory and office bulletin boards. Other scientists and academics joined in Ivanov's efforts. "When we arrived with spades, picks, and shovels, the constant sound of the artillery cannonade was still distant," recorded one employee of the Hermitage. "Then we became accustomed to the whine of shells, to nearby explosions."

As Ivanov worked on the fortifications, German planes would sometimes fly overhead. The Russians would dive for cover as the pilots strafed and dropped bombs. Nazi broadcasts had informed the women workers to wear white, promising that this way the bombers could avoid attacking them. Hundreds of naïve helpers wore white headscarves and shawls over their shoulders, only to be gunned down in the trenches, having made themselves easy targets.

In the park beside Pavlovsk Palace, where Ivanov and his wife had planned to walk on the afternoon of the invasion, a group of students headed toward the Luga Line bedded down for the night. Before anyone was able to fall asleep, German planes began bombing a nearby airfield. Soldiers ordered the students to hide anything that was white, extinguish cigarettes, and follow them to safety.

"We started walking fast along a road already full of our units," recalled Olga Grechina, a seventeen-year-old student at Leningrad University who was among the group. "The soldiers marched quickly and quietly; if one made a sound, the others shushed him for being careless. None of us had any idea where we were going or why, which made it even more frightening. We were all desperate for something to drink, so much so that when the road went through a wood, we drank muddy water from the roadside ditches."

The following morning, having marched nearly twenty miles, the group reached a village where they were handed shovels, axes, and stretchers for carrying soil and told to dig deep antitank ditches. After two weeks of this hot, wearying work, Grechina could no longer straighten her back.

Ivanov was used to toiling in fields, and the chance at last to do something practical assuaged some of the frustration of recent weeks. By the end of July the Luga Line was complete. The botanist returned to Leningrad with aching muscles. While information about the German advance remained scant, Ivanov could at least be sure enemy forces had not yet made it as far as the suburbs, even if their planes had.

In Leningrad Ivanov and his colleagues spent their evenings on the Institute's roof, where they could watch the beams of hundreds of searchlights cleave the sky, even if, in the early weeks of summer while the sun lingers late over Leningrad, they were of little practical use.

On July 18, as the second Hermitage train prepared to depart, the city's leaders introduced the first ration cards for bread, butter, and other essentials: four hundred grams (14 ounces) of bread, around six slices, per day, and six hundred grams (21 ounces) of butter per month. When one academic suggested that famine might come to the city, her friend scolded her, saying, "There is no reason to be pessimistic."

This optimism equated to mass delusion. Eleven days earlier, in Berlin, the Führer had met with General Franz Halder. Hitler told his military chief of staff that Leningrad was not merely to be attacked, but was to be leveled, to become "uninhabitable." This erasure was more symbolic than strategic. By razing the city the German army would eliminate a center of Bolshevism and nationalism. But there was a practical, if inhumane, benefit to Hitler's scheme: by reducing the number of people living in Leningrad the German army would be spared "the necessity of having to feed the population through the Winter."

The depth of the Führer's hatred toward the city was fully revealed when, several months later, he told his lunchtime guests, "The foundation of St. Petersburg . . . was a fatal event in the history of Europe; and it must therefore disappear utterly from the earth's surface."

As Army Group North advanced on the city and its people, the commander prepared his men to cast out all empathy and compassion for the enemy: "This war must have at its goal the destruction of today's Russia, and for this reason it must be conducted with unheard-of harshness. Every clash, from conception to execution, must be guided by an iron determination to annihilate the enemy completely and utterly."

Three million people remained in Leningrad, including several dozen employees of the Plant Institute. If Operation Barbarossa went to plan, by Christmas there would be none.

Officials eventually allocated to the Plant Institute two standard carriages and an open-top freight car for the evacuation of the seed—enough for just 4 percent of the collection—which was scheduled to take place on August 25, 1941. Each evacuee staff member was given an allocation of seeds to be placed in their hand luggage, a soft parcel weighing no more than five pounds.

IV

A TRAIN TO NOWHERE

I

THROUGH THE TRAIN-CARRIAGE WINDOW OLGA Voskresenskaya spied a small girl in a chintz dress lying on a hammock strung between two silver birch trees. The Soviet Union had been at war for nearly two months. But the passing scene of rustic normality restored to Voskresenskaya some calm. "Perhaps," she thought, "this will all pass." But as the train pulled into the scenic town of Pavlovsk on the outskirts of Leningrad, she was greeted with a forlorn scene, and the feeling dissipated.

Pavlovsk was a town of faded grandeur and old-world, cobblestone charm. The streets were lined with elegant mansions and stately trees. The air was perfumed with the scent of blooming flowers and freshly baked bread, and the sound of carriages clattering along the roads. The town's centerpiece was its grand palace, a majestic structure of white stone and gold leaf that shimmered in the sunlight, surrounded by equally magnificent gardens, with meandering pathways that wound through the trees and flower beds. It was a place to relax and recuperate; many of Leningrad's citizens would visit to unwind during holidays.

That day, however, Voskresenskaya and her colleague from the seed bank Abram Kameraz passed a stream of peasants leading creaking horse-drawn carts laden with their belongings. They were leaving. Children wept into the shoulders of the adults who carried them, while Soviet soldiers aimlessly looked on. In recent weeks Pavlovsk had transformed from a picture-postcard destination into a hub for refugees headed toward the city as they fled the incoming German army. More than two thousand evacuees from Pskov, from the Baltic, from Velikiye Luki and Gatchina, had taken up residence in the palace cellars, a temporary rest stop before they attempted to reach the city proper.

The town was heavy with the melancholy of people forced to relinquish their identities—postman, farmer, academic, parent—for a new and unwelcome label: refugee. The effort to leave one's home is as psychic as it is physical. Before the bombs fall and the flames rise, a new reality must be taken on faith. The future feels as illusory as a dream. Are they really coming? Must we really leave? Now?

Vavilov had instilled in his followers a keen sense of responsibility; many of the specimens in the seed bank, he taught them, were as irreplaceable as the artworks hanging in the Hermitage. They could not easily be re-collected or, in some cases, replaced at all, as the landscapes from which they had been harvested had already been destroyed by human activity. His staff understood that preserving the collection was now their primary goal. So, in mid-August 1941, eight weeks after the invasion began, absent government support, and with the enemy approaching the suburbs, the two potato experts had decided to act.

Kameraz and Voskresenskaya were friends and colleagues. Together with Olga's husband, Vadim Lekhnovich, and Fatikh Bakhteyev—the two men who had accompanied Vavilov on his fateful last mission to Ukraine—the pair represented some of the most talented and experienced tuber researchers in the world ("the whales on whose backs Soviet potato science rested," as one colleague described them).

In 1926 Vavilov had established a field station here in Pavlovsk, a place where Institute staff could grow and study perennial grasses, cruciferous tubers, fruits and berries, and potatoes away from the hubbub of the city center. Known as the Red Ploughman,* it was one of eleven

---

* *Krasny Pakhar:* Like many Soviet factories, products, and professions at the time, the *red* is a Communist epithet that refers to its Bolshevik associations.

satellite research stations that Vavilov established across the country, where specimens could be observed growing in different kinds of soils and climates.

A few months earlier, in the spring, Voskresenskaya and Kameraz had planted more than twelve hundred varieties of potato in the fields here at the Red Ploughman. Most of the Institute's botanical samples were stored in packets and filed for safekeeping in the wooden drawers that lined the seed bank's dozens of rooms. They could be kept in snug, dry compartments for years before they lost their latent, life-blooming potential. The potatoes were different, however. They had to be renewed and maintained through annual planting. And once planted, the tubers—the storage organs required to propagate new crops— needed time to mature before they could be dug up and preserved for another year.

Voskresenskaya cried when she and Kameraz arrived at the field station. A German shell had landed in one of the fields, scattering potato shoots like confetti. Some of the varieties were as small as pigeon eggs, and delicate to the touch. There was no chance they could survive an explosive blast. It had been an abundant year for the crop. But as the two colleagues pulled surviving plants from the soil, it was clear that many of them needed more time to grow—time the botanists could no longer spare.

The pair had another problem. The South American varieties were too delicate and underdeveloped to be transported yet. During the summer months, the sun sets on Leningrad after midnight and rises just two hours later. The sky remains pallid and milky throughout the night. The Chilean potatoes were used to shorter daylight hours so, under northern Russian skies, they had not yet produced tubers. To encourage growth, the scientists had artificially restricted sunlight using plywood cubicles that could be wheeled into place on wooden rails. Each cubicle had a double door that the botanists could open and close to throw the plants into darkness, a mechanism known as a photoperiodic plot. One hundred and twenty cubicles had been built by Institute staff, enough to cover the entire Latin American collection, which had been planted in pots, rather than in the ground. Despite the ingenious housing, the potatoes needed more time to mature. The botanists agreed to leave these specimens until the last possible moment.

For now, Voskresenskaya and Kameraz focused instead on the European varieties. The pair used forks and spades, carefully placing each plant in a collection box to be loaded onto one of the two waiting trucks bound for St. Isaac's Square. Kameraz had convinced the commander of a local unit of soldiers that was constructing a branch of the railroad for armored trains not far from the village of Pyazelevo to provide the botanists with the transport. The drivers would take the botanists and their harvest to the seed bank, then return to their unit from the city carrying ammunition and supplies.

A local farmer had volunteered to assist the botanists. As they carefully placed the specimens into burlap sacks, he congratulated Voskresenskaya on the size and number of tubers on each plant. His praise had calmed her, reminding her that their work was valuable, and the risks involved worth taking.

The group labored for hours until, when the trucks were full, the two botanists boarded and began the hour-long return journey to the Plant Institute, where Voskresenskaya's husband, Vadim Lekhnovich, had recently installed shelving to support the sacks in the cool dark of the cellar.

Voskresenskaya looked out onto the deserted countryside as the truck strained to ascend the Pulkovo Heights, a range of hills seven miles to Leningrad's south. The pale sky flashed intermittently with artillery fire. Then, from over the tree line, she spied a German plane flying in low and fast. The driver accelerated and, without turning, growled that, if they survived the next few moments, this would be the last trip he made to collect potatoes.

II

IN EARLY AUGUST, WHEN VOSKRESENSKAYA and Kameraz began their rescue mission, there was still a lack of clarity about the advance of Germany's Army Group North toward the city. Leningrad's citizens were left to disentangle fact from the euphemistic language regu-larly employed by Sovinform. Reports of *ozhestochenniye boi*, "bitter fighting," were less serious, it seemed, than reports of *uporniye boi*, "determined fighting." Both were preferable to *tyazheliye boi*, "heavy fighting," which, most people assumed, was code for "total annihila-tion." Few dared discuss Soviet losses, or the apparent failings of the city's leaders, in anything but whispers among trusted loved ones. As one university student put it, "There were always informants around."

In fact, after the fearfully quick early thrust through Lithuania and Latvia, Field Marshal Leeb's advance had slowed. The Luga Line, a 185-mile-long system of gorges, trenches, and "dragon's teeth"— concrete mini-pyramids designed to prevent the advance of tanks— had provided a firm buffer that stretched from the city of Luga toward Narva, near the Gulf of Finland and further slowed the Germans, who were already having to contend with challenging terrain, hot climate, and outdated maps.

Russia's more remote bridges were unable to sustain the weight of the German panzers, and roads the German commanders had as-sumed would be robust and maintained often turned out to be little more than dirt tracks. As one German officer put it, turning off one of the major roads was "like leaving the twentieth century for the Middle Ages."

Fortuitously for the Russians, summer thunderstorms began to turn dust to mud, stalling the trucks that carried the necessary supplies to keep the German tanks moving. This reprieve allowed Stalin's chief of staff, General Georgi Zhukov, to launch a counterattack against the Eighth Panzer Division, a skirmish that eliminated 70 of the group's 150 tanks.

On August 10, 1941, General Halder noted that Leeb's gains had been "very insignificant." The Luga Line had almost stalled the German advance. "What we are doing now," Halder continued, "is the last des-perate attempt to prevent our front line becoming frozen in positional warfare. . . . The critical situation makes it increasingly plain that we underestimated the Russian colossus."

The Red Army's commanders, meanwhile, had to contend with a kind of vindictive friendly fire, too: General Dmitri Pavlov of the Western Army Group was the first to be scapegoated by Moscow; he was arrested on July 4 and executed two weeks later, together with three of his subordinates. Lower-rank officers were routinely charged by military tribunals with cowardice and immediately shot. Certainly, there were deserters; in the week of August 16 more than four thousand servicemen were seized as suspected runaways trying to reach Leningrad; almost half were wounded in the left arm or hand, suspected of self-mutilation as a means of survival. On the outskirts of Leningrad, the city established a *komendatura* to check not only for German infiltrators disguised as refugees, but also for deserters.

Senior officers soon learned that surrender was preferable to retreat, which had become little more than a stay of execution. The captain of the troopship *Kazakhstan* was knocked unconscious and blown overboard in a bomb blast when the German army began to shell the Baltic fleet at Tallinn, the capital of Estonia. His limp body fell into the sea, where he was picked up by a passing Soviet submarine. Four days later, he was tried for "desertion" and executed by firing squad. The life-threatening incentive to hold the line, no matter the cost, was clear. So, too, was the cost to morale.

The Luga Line mostly held until August 8, when the Sixteenth Army outflanked it toward the south. The German troops had sustained losses while contending with fortified positions and tens of thousands of mines. Leeb now sought to rally his armies. He broadcast an optimistic message: "Soldiers! You see before you not only the remains of the Bolshevik Army but the last inhabitants of Leningrad. The city is empty. One last push and the Army Group Nord will celebrate victory! Soon the battle with Russia will be ended!"

With no reserves at their disposal, and no defensive lines remaining between the invaders and the city, the wearied officers of Leningrad Command were left with few options. Red Army troops were directed hither and thither to provide backup to beleaguered units. With each day that passed, evacuation from the city became more difficult and less likely. The direct Moscow–Leningrad railway track was blocked, so all trains were rerouted to the east, through the station town of Mga, forty miles from Leningrad, where the line split into two routes.

To the north of the city, the Finnish army had been advancing toward Leningrad, reclaiming territory taken by the Russians during the Winter War, two years earlier. Led by General Carl Mannerheim, Finnish troops had crossed into Russian Karelia and begun to advance along the northeastern shore of Lake Ladoga, the vast body of water that bordered the north of Leningrad's suburbs. Within days it seemed the Finnish troops would fulfill their promise to Hitler and "shake hands" with the Wehrmacht and form a circle around Leningrad. Then the only route into and out of Leningrad would be by boat or plane across the great lake.

III

As Voskresenskaya and Kameraz were working on the potatoes in Pavlovsk, the remaining seed bank staff focused on gathering valuable plant matter from the other satellite stations in the city's suburban areas in the direct path of the advancing German army. Ivanov helped the researcher V. F. Antropova evacuate pea and lupine samples from the Pushkin laboratories, along with a rare collection of rye seeds and melons, delivering them to Leningrad.

These measures provided the welcome distraction of activity, but as rumors trickled into the city of the Germans' swift advance, frustrations grew. Many of the seed bank's dozens of researchers, botanists, and administrative staff had been drafted into the army; others who were not called up volunteered for what became known as the *narodnoe opolcheniye*—the people's army, consisting of 130,000 men and women. They received no training and antique weapons. And still, through it all, no official word on what should happen to the seeds.

Two weeks later, on August 21, *Leningradskaya Pravda* lifted the veil of denial when it informed readers that the German armies might try to take Leningrad after all. The news of impending danger was met with widespread panic. Many had already survived starvation during the Revolution and Civil War era, when almost twenty thousand people had died in Leningrad. More recently still, Leningrad's citizens had contended with winter famines in 1936 and 1940. Survivors understood the life-sustaining importance of filling their cupboards and stockpiling supplies before the moment of disaster. Fearing a repeat of those shortages, hundreds of thousands of families scrambled to stockpile food. Crowds of citizens chased rumors around the city that this or that establishment was distributing supplies.

Only a relatively few nonworking people had been ordered to evacuate. Now union offices across the city were deluged with requests for passes to leave the city. The opportunity to escape had narrowed to a crack. So, block by block the city prepared for violence. Volunteers and soldiers dug thousands of miles of ditches and erected approximately fifteen thousand pillboxes—reinforced-concrete fortifications that provided shelter and protection for the soldiers and snipers inside—alongside fields of antitank mines. There were not enough guns to arm every able-bodied person, so those in charge of arming the defense units instead distributed long-bladed hunting weapons known as Finnish knives. On August 24, 1941, a curfew descended: all movement between the hours of 10:00 p.m. and 5:00 a.m. was prohibited. Nothing could be done about the lingering twilight of late summer in Leningrad. But in the early hours of the morning no lamplights or saboteur's signals could direct enemy bombers over the blacked-out, silent city.

For the botanists at the seed bank, clarity about the military situation was accompanied, finally, by the news that the Plant Institute was to be included in what would be the final set of industrial enterprises and institutes to be evacuated from the city. It was unclear if this represented a change in policy from Moscow, but whatever the reason a hundred Institute staff and their families would be taken by train along with the seed collection and relocated to Krasnoufimsk, a small town in the central Urals, just forty miles from the Hermitage's safely stowed treasures.

Any sense of elation was soon tempered, however, when representatives of the Oktyabrskaya Zheleznaya Doroga—the October

Railway—informed the seed-bank staff they had been allocated just two standard carriages and an open-top freight car for the entire collection. The staff felt anger and dismay: having patiently waited for weeks, the carriage would only provide room to evacuate a fragment of the Institute's 120 tons' worth of seeds.

IV

AS THE SCIENTISTS PREPARED THE seeds for their long-awaited evacuation, they were subjected to further needling and stress with the arrival of Trofim Lysenko's close collaborator, the thirty-eight-year-old Isaak Izrailevich Prezent, sent from Moscow to oversee the evacuation of the collection.

Prezent was a little man with large aspirations. He wore shoes with built-up heels and a tall green hat. One colleague at Moscow University described him as a "strange, short man," difficult to like and a widely known harasser of women (he had once been expelled from Moscow University for attempting to seduce unreceptive students). He was also wily and ambitious. Years earlier he had briefly worked at the Plant Institute, but Vavilov had little time for this self-described "philologist," whom the director soon judged to be a political chancer and impostor. The pair fell out. Prezent left the seed bank nursing a searing grudge.

Prezent and his geneticist wife, Basia Potashnikova, became vociferous critics of Vavilov and his staff, whom the couple both envied and feared as a potentially dangerous collective. "Vavilov's views obstruct our attempts to make the Institute of Plant Breeding more involved in the socialist construction," Potashnikova wrote. "It is very difficult to fight against them, as they have a very tight bunch of specialists."

Then, in 1932, Prezent met the peasant-agronomist Lysenko. He spied an opportunity to deal a blow to Vavilov's project and reputation. Prezent began to mold this promising but unpolished scientist into a star of his own design. Prezent coauthored articles with Lysenko and coedited the journal *Vernalization*, which dismissed the entire field of genetics as a "pseudoscience." Prezent helped Lysenko organize vast experiments and publish the dubious results and crafted compelling statements designed to uplift Lysenkoism, denigrate his opponents, and propel his protégé from the sphere of science to that of politics—to each man's mutual benefit.

Prezent had one especially useful talent in the context of Stalin's murderous, paranoid regime. As the novelist Vladimir Nabokov put it, "The successful Soviet writer was the one whose fine ear caught the soft whisper of an official suggestion long before it had become a blare." Prezent had a preternatural talent for articulating precisely what Stalin wanted to hear, and in Lysenko, Prezent had found an ideal sock puppet. Any doubt as to the source of Lysenko's attacks on Vavilov was settled when the agronomist admitted, "I simply get on with my work; Prezent works out the philosophy for me."

To Vavilov's staff it seemed clear that Prezent was somehow behind the suspicious circumstances of their leader's disappearance, and the terrible silence that had followed. The Institute staff had become used to endless demoralizing conflicts and inspections. During the past ten years, some of Vavilov's critics had been appointed to chair research units inside the Plant Institute and had used the opportunity to install a fifth column to work against its director. So when Prezent arrived to oversee the long-anticipated evacuation, Vavilov's supporters responded with cool, familiar resentment. Here was the man whose sly manipulation had led the pliable Lysenko to dismiss the Plant Institute as a center of rotten, spurious Western biology, and to scandalously describe Vavilov's work as "racist."

It was, however, a mercifully short-lived imposition. Either sensing the hostility of the group he encountered on arrival or concerned that he might become trapped in the imperiled city, the self-preserving Prezent left Leningrad almost as soon as he had arrived, ensuring that he would not have to defend his rival and enemy's seed bank if the Germans surrounded Leningrad.

Once Prezent had departed, the seed bank's acting director, Eichfeld, called Ivanov and other senior staff to his office to manage preparations. The freight train, he explained, was scheduled to depart imminently, on August 25, 1941. This left little time to select which of the hundreds of thousands of specimens to remove from the city.

"We'll have to work around the clock," Eichfeld despaired.

The Institute's apple expert, Rudolf Yanovich Kordon, a Latvian with a Cossack-style mustache, suggested the botanists select only duplicate seeds for transportation. "That way, if something happens to the train, we have reserve stock," he said.

Ivanov agreed. "We should not take risks."

There was no chance of securing additional freight carriages. The Plant Institute would share the limited space on the train with the other institutions, each of which viewed its collections with a similar sense of priority and concern. To increase the number of seeds they could evacuate, the group devised a plan: each passenger would be given an allocation of seeds to be placed in their hand luggage, a soft parcel weighing no more than five pounds.

These parcels would contain one of a hundred cereal grains—wheat, rye, barley, groundnuts—and between fifty and a hundred other assorted crop seeds. In this way, a further twenty thousand samples could be added to those held in the freight car. Employees would have to make room in their luggage for the parcel, but no one quibbled; the seeds were of greater value than other belongings.

For the freight car, Ivanov and his colleagues filled double-walled boxes with around one hundred thousand seeds, as well as valuable scientific equipment, and rare books and manuscripts from the library. Each seed variety was allocated a number and supporting documentation, to ease the sorting of the material when it arrived at its destination. The combined weight of the material amounted to five metric tonnes, equal to that of a fully grown elephant, but represented just 4 percent of the Institute's full collection.

Not every seed bank employee accepted this eleventh-hour chance to escape. While plans were being finalized, Ivanov met Aleksandr Gavrilovich Shchukin, the Institute's groundnut expert, in the corridor.

"They want to uproot me, at my age?" Shchukin exclaimed to Ivanov, his meticulously kept log of groundnuts under his arm.

For some of the older employees, the risks of staying in the familiar city were still not grave enough to inspire escape. Ivanov loved Leningrad, too, not only for its majestic spaces and grandiose monuments, but also its more liminal qualities: its intellectual endeavor, ethereal intensity, and capacity to inspire poets and painters. In what other city would the scholar eagerly break bread with the artist, the composer with the priest, the ballet dancer with the botanist?

At home that night Ivanov discussed the matter with his wife. The pair decided they would stay. Ivanov would remain with the seed bank and guard the remnant. The next day he offered his seat on the train to someone else.

When the collection was packaged and ready, soldiers collected the crates from the seed bank and escorted them to the railway depot, to be loaded onto the waiting train. Dozens of representatives of other institutions in the city had come to load their documents and materials as well. Three members of the seed bank staff stayed with the collection and took turns to guard the carriage on the siding overnight, lest thieves attempt to break in and use the collection to fill their cupboards and larders.

V

ON THE MORNING OF AUGUST 25 the seed bank staff who had elected to remain in the city made their farewells to those who had chosen to flee. The acting director, Eichfeld, signed an order passing leadership of the Plant Institute and its field stations in the suburbs at Pushkin and Pavlovsk to his colleague Yan Virs. Then Eichfeld and his deputy, G. G. Kovyazin, set out for the station.

Before the staff and collection could set off, another train filled with evacuee children had to leave first and clear the line. The Germans had destroyed the bridge at Volkhovstroi, en route to Vologda, leaving just one track open, the Pestovo line. The train filled with children set off early. It would have to stop at the transfer station of Mga, to collect more evacuees, before heading farther east. As the driver began to apply the brakes, a boy pointed out the window of one of the carriages, telling his friends to look at the balloon floating in the sky. Except it wasn't a balloon. The child had spotted the parachute of a German paratrooper drifting into a field on the outskirts of town.

All train drivers knew of the risks in stopping to collect more passengers before they were clear of the suburbs. The previous month a train carrying around forty children out of the city had been targeted by German planes. While helpers lifted the children's luggage aboard at Lychkovo, enemy planes flew low overhead; some dropped bombs, other raked the train with machine-gun fire. As the carriages filled with smoke, chaperones placed mattresses over the children's bodies to protect them. Then there had been a huge explosion. When the smoke cleared, carriages were scattered everywhere, as if, one eyewitness recorded, they had been "knocked off the track by a giant hand." Survivors saw tiny limbs caught in overheard wires and tree branches.

This tragedy must have been sharp in the driver's memory on August 25, as he saw paratroopers drifting into the city, and accelerated accordingly. Around a thousand children stood on the platform as their train out of the city streaked past. They had missed their chance to escape Leningrad.

The battle for Mga delayed the botanist's train for several days. Perhaps the evacuee botanists knew this might be the last opportunity to remove a trainload of valuables from the city when word reached them that it was time to leave. The fighting was still intense[*] when the train carrying the seed bank staff and some of their collection departed Moskovskaya-Sortirovochnaya Station in the southeast of the city. The passengers felt the sway of the carriage as the driver periodically accelerated, then braked, for unknown reasons, while following the Neva River eastward.

It had been a dry, warm summer in the city, and the trees had held their leaves, a vibrant display of russet colors: amber, yellow, scarlet.

---

[*] The hitherto obscure Mga was captured on August 29, 1941. The following day an NKVD rifle division recaptured the town. The German Twentieth Motorized Corps countered and, on August 30, retook Mga for the final time that year, facilitating, a few days later, the beginning of the Leningrad blockade.

On the outskirts of Leningrad, however, the countryside already bore the pocks of war: craters on either side of the tracks filled with water, splintered telegraph lines, earth scorched by explosives, trees snapped and uprooted. As she looked at a snaggle of roots and broken branches through the window, the writer Inber noted, "Here is the whole history of [the tree's] life, and now this history is cut short in the middle of a sentence. Everything is split, charred, dead."

Wearied from the busyness and stress of the past week, the seed bank's staff nodded off, surrounded by luggage containing soft packets of seeds drawn from around the world.

VI

A FEW DAYS BEFORE THE train set off, Joseph Goebbels, Germany's minister of propaganda, visited Hitler at the Wolf's Lair, a camouflaged and reinforced bunker hidden in the Polish woods, which served as the Nazis' Eastern Front headquarters. Hitler had just recovered from a bout of dysentery. The pair strolled in the woods and talked for four hours. Hitler explained that his plan to delay the army's advance into mainland Russia until the fall of Leningrad had changed. The advance of Army Group Center along the Moscow axis was now the priority.

What will happen to Leningrad? Goebbels asked. No longer would the German army take Leningrad by force, Hitler explained. Instead, they would surround the city and "starve it into submission." And once food and supplies could no longer enter the city by road or rail, the Luftwaffe would begin a bombing campaign to hasten the citywide famine, and target the city's water, power, and gas stations.

News reached the front line quickly. On August 29 Leeb ordered his forces to close the ring around the city. The general was desperate to complete this part of the mission before any of his units were ordered south to join the assault on Moscow. The following day, German forces captured Mga, and with it closed the final rail link from Leningrad.

At dawn that morning, evacuees from the seed bank, whose

journey had been slow and faltering, were awoken by a sharp jolt. Some passengers opened the carriage doors and began to walk along the track toward the front of the train to find out what was happening. A Red Army sergeant, his bandage soaked with blood, was addressing the driver.

"The Nazis have taken Mga," the sergeant said. "The track is blocked."

More than two thousand goods vans and cars loaded with the property of many of the city's enterprises and offices cluttered the tracks, unable to pass through Mga into the Russian hinterlands. Unshaved, with sunken eyes, the officer could barely hold himself upright. Only fifteen men from his unit remained, he explained.

"Turn around and take us with you," he pleaded.

The evacuation had failed. The Institute staff and their families would have to return to Leningrad and leave the open freight carriage containing the double-walled boxes of seed samples and important documents forsaken on the siding there.

Throughout late summer 1941, Abram Kameraz and his colleague Olga Voskresenskaya fetched potatoes from the seed bank's field station in Pavlovsk, a town in Leningrad's suburbs. Kameraz was digging South American varieties of potato when a German shell exploded nearby, knocking him unconscious to the ground. When he came to, he escaped the town and brought the specimens to Leningrad in a sack slung over his shoulder.

V

NO POTATO LEFT BEHIND

I

SOMEWHERE HIGH ABOVE, THE MOSQUITO drone of a plane's propeller neared. Abram Yakovlevich Kameraz was unconcerned. Since the thirty-six-year-old botanist had begun to commute by train to Pavlovsk, attacks by enemy planes had become as routine a cause of delays as leaves on the line. The local train no longer kept to a timetable. It ran only when it seemed safe to do so. Kameraz was one of the few passengers aboard. There were no good reasons for a civilian to make such a hazardous journey into the suburbs, toward the ranks of the rapidly advancing German army.

Throughout August, Kameraz and his colleague Olga Voskresen-skaya had made regular trips back and forth between Leningrad and Pavlovsk. But after the enemy planes fired on the trucks carrying potatoes near the Pulkovo Heights, the military drivers had refused to take them. The local commander explained that if Kameraz wanted to save any more potatoes, he would have to travel by train.

On August 28, 1941, two days before Mga fell, the Germans began to shell the southern and southeastern suburbs of Leningrad from their position at Tosno, twenty miles away. Pavlovsk was a key target for the German offense. Its gardens overlooked the surrounding countryside; its adjacent fields—the same fields where the rare varieties of South American potato were growing in their enclosures—presented ideal locations for heavy artillery to shell the city as the siege ring closed in.

Through the carriage window, Kameraz saw that the road from Pavlovsk to Leningrad was littered with bodies. These men, women, and children had been killed by German planes that had strafed and bombed the crowds of refugees as they raced toward the city. The train, with its weak-walled carriages and predictable, track-borne route, was especially vulnerable to attack from the air. As Kameraz caught the silhouette of a German Stuka cresting the horizon, the driver applied the brakes and ordered everyone out to run for cover.

Kameraz made a brisk, cowed dash across the rain-drenched ground into a nearby ditch. He pressed his body flat and listened. The air carried a faint smell of smoke drifting from smoldering peat bogs the Russians had lit to confuse the enemy pilots. Kameraz knew from experience that the plane would probably pass by and ignore the train to reconnoiter the roads leading toward Leningrad. Perhaps today the pilot would dip and open fire, the bullets kicking up plumes of dust, or pocking holes into carriage roofs. To any sheltering passengers these attacks were frightening. Soon enough they would seem like childish salvos, incomparable to the apocalyptic violence destined for the city, and its gaunt, horrified people.

II

ON SEPTEMBER 5, 1941, AS lines of hopeful, desperate people stood by the shuttered ticket office windows at the city's train stations, hoping that the Red Army might retake Mga and restore the line to Moscow, German planes dropped thousands of propaganda leaflets onto Leningrad's streets. Each fluttering flyer carried an ultimatum: there would be a massive bombardment if the city's leaders did not immediately surrender. Most people left the papers wherever they landed, fearing reprisals should an informant spot them picking one up. The Germans' message soon spread nonetheless, via word of mouth. The pamphlets were addressed to women and urged mothers and wives to pressure their husbands and sons to lay down their arms. "Take every opportunity to convince [them] of the senselessness of struggling against the

German army," they read. "Only by ending the battle of Leningrad can you save your lives."

The city was all but surrounded. To the north, the German's Finnish allies had made steady progress and had reached the northern shores of Lake Ladoga. When a group of twenty-eight young Communists were captured while defending an airfield outside Vyborg, an officer told them, in Russian, "The valiant German troops already are marching down the boulevards of Leningrad. And you stupid kids think you are going to save Russia. What's the idea of this suicide? I ask you. Soviet Russia is kaput."

The long summer days had begun to shrink, taking with them the flimsy comfort of ignorance. Throughout July and August no bombs had fallen on Leningrad, which inspired a rumor that Hitler had a secret daughter who wanted the city left intact for her future residence. Now that delusion was shattered. The fall of Mga, formerly a peripheral pass-through and now the focal point of a million hopes and anxieties, had not only halted the evacuation of Leningrad, but had also brought German artillery into range of the city center. The first shells landed on September 4.

A German plane was yet to bomb the city, but air-raid alarms now sounded regularly. The wailing heightened anxieties. "We feel the Junkers circle the outskirts of the city but cannot break through . . . but the day will come when they will." For those who felt abandoned by Moscow or enthralled by the gathering display of German might, the promise of the invader's imminent arrival was cause for tentative celebration, not dismay. The German army would come, they hoped, not as vanquishers but as saviors.

Three days later, as the sun ripened into its evening hues, the composer Gavriil Popov sat down at the piano in the apartment owned by his friend, the artist and founder of the Leningrad Puppet Theater, Lyubov Vasilievna Shaporina. Popov called to Shaporina to join him while he played Modest Mussorgsky's "Promenade." The apartment, where Shaporina's husband, the composer Yury Shaporin, would write and play, often resounded with music. But to perform a piece for one's friends while the city braced for impact recast entertainment as an act of defiance.

Shaporina listened as the musician's fingers quickened during the most virtuoso passage. As the piece began to crescendo, Popov stopped and exclaimed. "They're shooting!" The pair raced to the window. White balls exploded overhead, like dandelion puffs pressed into the sky by Leningrad's antiaircraft guns. After days of softening the city with artillery attacks, the German bombers had arrived.

Barrage balloons hung in the skies above the seed bank. The cables that tethered them to the ground were designed to catch the wings of bombers and send them into unrecoverable spirals. Popov and Shaporina watched as the Luftwaffe's pilots maneuvered around the cables with ease. "They might as well be soap bubbles," wrote one onlooker. Throughout the summer, Leningrad's citizens had grown used to air-raid sirens that came to nothing. Today, however, the throb of the twenty-seven Junker bombers provided a new sensation, as did the deep-thudding detonation of their bombs, first heard and then felt.

"It sounded like the sky was falling down," wrote one sixteen-year-old schoolgirl, who kept her clothes on when she went to bed that night in case she needed to run for shelter. Explosion followed explosion, the dread timpani of violence, each blast another crater in the lives of the loved ones of those it obliterated.

At the Musical Comedy Theater on Arts Square, audience members were enjoying the interval of a performance of Strauss's *Die Fledermaus* when the manager entered the foyer and told his patrons to press themselves against the walls. Gesturing toward the theater's domed ceiling, the theater manager pointed out that there was little protection overhead. The audience waited for forty minutes till the all clear sounded, then they returned to their seats and the operetta continued—albeit with each piece played at an increased tempo, and the less important arias and duets omitted. Most understood it to have been another routine drill, until they emerged, blinking, at the end of the performance and stepped into a charred night. "Suddenly we saw black, swirling mountains of smoke, illuminated from below by flames," recorded one audience-member. "All hell was let loose in the sky."

From behind the roofline, Popov and Shaporina saw a vast white cloud begin to rise on the horizon. It grew quickly, arcing, curving, tumbling outward, until the sky was obscured by gray clouds, tinged with amber from the setting sun. The pair watched in confused awe as a black stripe began to creep upward in the sky. The Junkers had hit their target: the so-called Badayev warehouses situated on Kievskaya Street, named after the Soviet politician Alexei Badayev. Comprising thirty-eight wooden warehouses close to the Warsaw railway station, it was here that Leningrad's leaders had concentrated a considerable propor-tion of the city's remaining food stocks. The flammable buildings, po-sitioned no more than thirty feet apart from one another, housed three thousand tons of flour and two and a half thousand tons of sugar; after the first bomb struck, the flames leaped from warehouse to warehouse. The blaze soon covered four acres, a fire of inconceivable intensity.

"It was so unlike smoke," Shaporina wrote in her diary that night, "that for a long time I could not comprehend that it was fire. . . . It was an immense spectacle of stunning beauty."

By the end of the night Leningrad's food stocks lay in piles of smoldering ash that coated the warehouse district. In weeks to come scavengers would come here and scoop the ashes from the scorched ground, sifting for teardrops of caramelized sugar to feed themselves and their children.

III

BY THE TIME KAMERAZ RAN for the ditch beside the railway line, the monkeys in the zoo had become so traumatized from the German shelling that they would sit in unblinking silence during raids, impas-sive toward the sound of each explosion. Despite the risk in journeying to Pavlovsk, Kameraz had been unable to push thoughts of the fragile South American varieties of potato in their cubicles from his thoughts. September had brought with it unusually early frosts, which threat-ened the crop.

Even as German troops had begun their pincer movement on the town in early September—one group advancing along the road to Pavlovsk and Pushkin, the other along the road to Krasny Bor, Yam-Izhora, and Kolpino—he had decided to stage today's reckless final rescue attempt. Every potato saved increased the chances of preserving not only his research work, but also the physical objects that Vavilov had carried across the world. The vegetables provided a sharply meaningful link between the efforts of the past, and the hopes for the future.

Kameraz was used to crouching in ditches, but as a botanist not a soldier. He had grown up poor on the banks of the Okhta River, east of Leningrad. When the Civil War had brought famine to the household, he borrowed books on agriculture from the local library and taught himself how to grow potatoes and pumpkins on a patch of nearby land. This humble allotment soon provided vegetables to feed the family.

In 1924, when he was eighteen, Kameraz enrolled at the newly formed Leningrad Agricultural Institute. There he attended a lecture by Vavilov, the newly installed director of the Plant Institute in Leningrad. The charismatic speaker had recently returned from North America and spoke vividly of his expeditions to exotic countries in search of rare seeds. Vavilov left an indelible impression on Kameraz, who applied to work for him immediately after his graduation in 1927.

At twenty-three Kameraz joined the seed bank as an intern, then worked his way up from lab technician to senior research officer. In 1936 he was awarded a PhD for his work on the properties of South American potatoes. Then, a few months before the invasion, he became the Plant Institute's potato specialist, a maven of tubers. Kameraz was responsible for overseeing the care and cultivation of the rare potato samples that Vavilov and his teams brought to Russia from their expeditions.

Under Vavilov's direction, the Plant Institute's potato collection had grown to six thousand varieties, including almost every breed cultivated worldwide, alongside wild types and new cultivars discovered and collected during expeditions to Latin America. Kameraz's task was to experiment with potato crossbreeding in the hope he might develop resilient hybrids. It was the largest, most diverse collection of potatoes ever gathered in history, a crop of inestimable scientific importance.

From his position in the trench, Kameraz heard the train driver's "All clear." He stood to his feet, dusted his clothes, and clambered onto the siding to retake his seat in the carriage. Through the rain-grimed window, Kameraz watched as the train passed into dacha country, the suburbs where wealthy Leningraders kept second homes for country-side retreats. The farther the train traveled from the city, the more deserted the roads. He was headed into a battlefield.

The train pulled into an empty station in the early evening. Kameraz disembarked into a town filled with overwhelming silence and gathering dusk. Pavlovsk's residents and refugees had fled, leaving the town to soldiers. Through empty streets he made for the Plant Institute's field station. On his way he ran into a Russian patrol. The soldiers demanded to see his papers. Kameraz presented the permit he had recently been given by the chairman of the Slutsk Executive Committee, the official name for Pavlovsk's town council at that time, a man named Korovichev. A few weeks earlier Korovichev had been astounded by the botanist's appeal to continue his work in Pavlovsk. The skies were darkened by artillery fire, roads were cratered from falling explosives, and this scientist wanted to dig up vegetables?

"Who cares about potatoes?" Korovichev had told Kameraz. "This is a war zone."

Kameraz had insisted that the collection was of "enormous significance," not only to the scientists who worked at the Institute, but also the farmers who would eventually grow these crops to feed the people of Russia. With the memory of famine still close in the national memory, no official wanted to be responsible for stifling food production—even on a front line. So Korovichev had provided Kameraz with a signed safe-conduct pass that allowed the botanist free passage into the town, and, where possible, military assistance. Now Kameraz presented this document to the soldiers. The patrol waved him on.

The Red Ploughman field station was abandoned when Kameraz arrived. The laboratory technicians and farmworkers who had assisted with the evacuation of plants throughout August had fled, leaving the plants untended in their wooden cubicles. Kameraz opened the sliding doors on one of the little sheds to allow the light in. Then, one by one, he took each Chilean plant from its pot and gently tapped the soil free, checking to see which specimens could bear the stress of transport. Kameraz's practiced hands worked quickly. He labeled any plant whose tuber had sprouted, wrapped it in parcel paper, and placed it into a sack, ready to be slung on his shoulder for the return journey to Leningrad.

Throughout the week he had been able to continue this solitary work undisturbed, carrying two full sacks home each day. Today, however, the atmosphere in the town was different. Kameraz heard the pock and shudder of nearby artillery fire. The German forces were even closer than the Soviet authorities had been willing to admit during their five daily announcements, broadcast through the city's expansive network of loudspeakers. In fact, at that moment the Russian 260th Regiment had retreated so far that its command post was now only half a mile from the Pavlovsk Palace gardens. The forward edge of the battle was a ten-minute walk from the town center. After weeks of wondering, Kameraz realized the invaders had arrived.

He took cover inside one of the sheds. As the shelling intensified, and the sound of explosions drew closer, Kameraz wondered if these were to be his last moments. Shells had hit the potato plot before. There was a good chance they would do so again. After a while, however, he grew accustomed to the noise of proximate violence. He tuned the sound out, opened the shed door, checked the tree line, then gingerly resumed his work. He moved carefully from shed to shed, checking the plants and bagging those he considered mature enough to survive the journey home.

Then there was a sudden flash, followed by black, ineffable silence.

IV

SEVERAL HOURS BEFORE THE BLITZ of the city center began on September 8, those seed bank staff who had failed to evacuate returned to

Leningrad, their luggage still filled with the carefully wrapped packages of seeds. After they had disembarked, the open freight carriage laden with boxes of seed samples had been pulled along the line from the Moskovskaya-Sortirovochnaya Station in the south to Toksovo in the north and placed on a siding, outside the target area of the Luftwaffe's bombing runs. The plan was simple: as soon as a train route out of the city reopened, the collection could resume its evacuation journey.

After a sorrowful reunion with their colleagues, around thirty staff members moved into the seed bank's buildings, which were now braced for the onslaught from the skies. The local police department issued a single nighttime pass to the Institute, which was claimed by the building manager, Maria Belyaeva. Her colleagues would have to remain indoors during curfew hours. Leningrad's officials and residents alike knew of the Nazi blitz of London, which had ruined the city and nightly terrified its citizens with oil bombs dropped from planes that doused streets in flaming liquid. Just as Leningrad's residents were ordered to remove sheets, old newspapers, boxes, and other flammable objects and materials from their lofts and attic rooms, so Ivanov and his colleagues began to move archival manuscripts and chemicals to the basement.

The Leningrad city council had ordered the creation of ten thousand firefighting units, many composed of civilian volunteers, and assigned each a factory, office block, department store, or set of apartments to protect. From his fire-watch post on the roof of the conservatory where he once studied, the composer Dmitri Shostakovich worked on a symphony he hoped would, in time, rouse the city's population. Half a mile away a firefighting group led by I. Ya. Yurtsev was allocated to the main seed bank buildings on Herzen Street. The Forest Aviation Trust requisitioned one of the rooms and the firefighters blacked out the windows.

Across Leningrad, the fire department built concrete water-storage basins and installed five hundred new water hydrants. The training offered to volunteer firefighters was superficial, and the equipment rudimentary: helmets, axes, crowbars, and buckets. As in many areas of civil defense, responsibility for fire watch often fell to women, who were shown how to use sand to douse an incendiary bomb, a narrow, flanged cylinder that smoldered on impact like a dropped match. One volunteer replied sarcastically while undergoing civil defense training, "And what if it is an explosive bomb?"

On the night of the attack on the Badayev warehouses, almost all the firefighting units were summoned to fight the blaze. Then, three and a half hours after the first attack, German planes began a second run, dropping forty-eight high-explosive bombs. One burst near the bridge crossing the Winter Canal, shattering windows throughout the Hermitage Museum, across from the seed bank, as the building sustained its first damage from German attack. While it was not a legitimate military target, Nazi forces had designated the Hermitage as Target Number Nine on their maps of Leningrad. A sign featuring a large arrow and the text BOMB SHELTER FOR PASSERSBY hung on the gates of the Hermitage. Anyone caught in the square during the raid could shelter in the museum's basement.

During the second attack, twenty-four people died inside a pumping plant at the city waterworks. Seventy animals at the zoo perished, including its most famous occupant, a fifty-year-old Asian elephant named Betty, who was crushed in her enclosure by falling rubble. The blaze at Badayev lasted throughout the night, an attack that many citizens would soon blame for the shortage of food.

Ships and harbor buildings became the focal point of the airborne attacks, as did the Kirov Plant, or Factory No. 100, first established 140 years earlier for the manufacture of cannonballs, which now manufactured T-34 tanks. In cellars, children wept as they pressed their heads inside their mothers' coats. With each new fly past and its trailing chain of explosions, women crossed themselves and whispered prayers.

"The din intensified," one resident wrote in her diary that night. "Sheer hell."

From the windows of the Plant Institute offices, Ivanov and his colleagues watched the flames rising from the factory. As acrid smoke swept through the streets, it stung their eyes and stuck in their throats. To the east, the town of Shlisselburg fell and with it the last roads into and out of the city. Leningrad had been cleaved from the *bolshaya zemlya*, the Russian mainland. Now supplies could only enter via plane or boat across the inconceivably vast Lake Ladoga to the northeast, an expanse of water large enough to fit fifteen cities the size of Los Angeles.

For now, the seed bank staff and most of the 2.5 million other trapped people would mostly have to survive on whatever food was already inside Leningrad's perimeter. Prior to the invasion, Russia had been energetically exporting food to Germany as per the terms of the nonaggression pact between the two countries; stocks were now so low that, by some estimates, there was less than one month's supply of many types of food.

In the diaries kept by ordinary Leningraders, two metaphors dominated: the city as an "island," the invaders as a "ring." Leningrad was isolated and surrounded. The poet Olga Bergholz chose a third image to describe the situation: "The blockade," she wrote, "tightened around the city's throat like a noose."

The morning after the first night of the blitz, rescuers tilled the rubble-like fields. The botanists stepped into streets that sparkled with broken glass. An editorial in *Leningradskaya Pravda* clarified to every reader the desperation of the situation: "The enemy is at the gates."

The men and women of Leningrad were no longer mere citizens; they were *blokadniki* and *blokadnitsy*: "blockade dwellers." Bergholz captured the transition in a few lines of melancholic, portentous verse:

> Leningrad in September, Leningrad in September,
> Golden twilight, the regal fall of the leaves,
> The crunch of the first bombs, the sob of the sirens,
> The dark and rusty contour of the barricades.

The siege had begun.

On September 4, 1941, Hitler's troops began to shell and bombard Leningrad for up to eighteen hours at a time. Staff at the seed bank recorded the attacks in a logbook. In total, 108 incendiaries landed on the Institute roof, which sustained the most damage during the night of October 1, when the botanists extinguished five fires.

VI

CITY OF FIRE

I

I𝐓 𝐰𝐚𝐬 𝐩𝐚𝐬𝐭 𝐦𝐢𝐝𝐧𝐢𝐠𝐡𝐭 𝐰𝐡𝐞𝐧 Abram Kameraz opened his eyes and, through a veil of concussion, saw the stars glinting overhead. The field was quiet, but the sound of a thousand crickets deafened the botanist. If he was in peril, there was little that he could do while disoriented and immobile on the soil. Kameraz attempted to lift himself onto his elbows to check for injuries. As he strained his neck, he was overcome by a surge of nausea and again blacked out.

Pavlovsk was deserted. The windows of the grand palace were boarded up. Inside, a few remaining staff worked to conceal its treasures in the cellars behind false brick walls by the light of burning ropes or twisted pieces of paper. The sculptures in the surrounding palace gardens had been buried, the disturbed earth padded down and covered with fallen leaves. On one statue a worker had written the words, "We'll come back for you." Into this same ground in Pavlovsk Park, the Nazi invaders would later bury at least forty-one executed Jews.

For a second time Kameraz came to, now in the cool morning light. He first checked his body for injury, then, when the memory of his mission broke through, checked the spilled sack to ensure the potato specimens had not been damaged in the blast. Finally, the botanist stood to his feet, hoisted the sack onto his shoulders, and made his way to the railway station, hoping there might still be a train to return him and his final haul to the city center.

When Kameraz finally reached the Plant Institute, he received a hero's welcome. He and Voskresenskaya had succeeded in delivering more than a metric tonne of potatoes, at least one specimen of every variety held in the Institute's collection. The botanists were used to considering the best conditions for ensuring the survival of their specimens. Fearing that their work might be obliterated in the blast of an explosive shell or burned to ash by incendiaries, the pair divided the potatoes into three duplicate sets. One tuber of each variety was placed into a drawer in the building at 44 Herzen Street; another two were moved into the cellar of 42 Herzen Street, where Voskresenskaya's husband had mounted shelving onto the walls. If bombs or fire destroyed one crop, the remainder would hopefully survive.

Pavlovsk and Pushkin had fallen. There would be no more trips to the Institute's two closest field stations. Already food in the city was becoming scarcer. Realizing that a cellar filled with potatoes might become a draw for citizens running low on supplies, Kameraz and his pregnant wife, Lyudmila, moved in to guard the store. The couple slept on the floor, using sacks for blankets. A new watch post was set up by the entrance to the basement. No one could enter the potato vault without permission.

II

THROUGHOUT SEPTEMBER THE PLANT INSTITUTE became something between a barracks and a closed society. Staff members slept on the premises and organized round-the-clock watches. Senior members of the team were appointed new roles and responsibilities. The Institute's head of personnel, Georgi Reuter, became secretary of party organization, and Dmitri Ivanov, rice specialist and head of the department of cereal crops, was promoted to chief of staff. The newly appointed on-premises director, Yan Virs, became the chief officer of the facility, while Nikolai Likhvonen, the seed bank's procurement agent, ran the so-called self-defense group, which was divided into several subunits, including groups for observation, communication, medical care, sanitation, firefighting, and repairs. Soldiers removed typewriters, medical equipment, and the telephone switchboard from the building for military use.

The evacuation of the seeds had failed. Now, as the enemy bombarded the city, the botanists took simple measures to improve security should the siege turn into a ransacking. Room 12 on the second floor of the building* at 44 Herzen Street became the assigned office for the seed bank's on-duty overseer. Inside, staff installed a cast-iron stove, along with a loudspeaker and a telephone. They arranged two folding beds and ten chairs behind a screen partition in the room, for when the occupants needed some rest. The duty officer was tasked with maintaining a daily record of the forty rooms where the bulk of the Institute's specimens not currently packed in boxes on a static freight carriage had been arranged. Each room was locked and stamped with a wax seal to show whether anyone had attempted to enter it unsupervised.

Staff could only enter the seed rooms once a week, and only in groups of three or four while accompanied by the on-duty overseer, and the chief keeper of the collection, Rudolf Kordon. Together they would check the condition of the boxes. Kordon was one of Vavilov's oldest employees, having joined the Plant Institute in 1926 after he graduated from Leningrad State University. He was the seed bank's resident apple expert and had before the war cultivated a variety of the fruit that bore his name, the Kordonovka.

Now Kordon worked closely with his friend and supervisor Klavdiya Panteleyeva, an energetic and efficient agronomist, who had propagated perennial flowers of the aster family before the war. The keys to the collection rooms were held in a safe, accessible only to Panteleyeva, to whom ultimate responsibility for the collection's safety now fell. After each daily inspection, the on-duty manager would write a brisk confirmation in the logbook: "Seals and locks intact."

These basic precautions would do nothing to hold back a determined enemy unit. But the administrative routine gave the staff

---

* Now part of the Department of Oat, Rye, and Barley Genetic Resources.

a schedule and a purpose and, if food became scarcer still, provided some collective accountability to dissuade anyone from the temptation to supplement their rations with seeds and nuts from the collection.

Staff also recorded in the logbook a time stamp to mark the sounding of every air-raid alarm and all-clear signal, a record of any proximate bomb blasts, and whenever an incendiary bomb landed on the roof. German bombing runs were regretfully frequent. Young musicians from the Leningrad Conservatory would stand among the columns at St. Isaac's Cathedral, close to the seed bank, to listen for the approach of engines. Believed to have superior hearing, the students would telephone the on-duty air defense officer with information about the direction of approach before the planes were sighted by lookouts—though this inventive scheme did little to mitigate the air assault.

The seed bank staff would hear the whining drone of a solitary German Heinkel bomber before they saw it. Its appearance soon became routine and feared, and to Ivanov and his colleagues it seemed the Luftwaffe was focused on their priceless collection of seeds. At night a pilot would drop a parachute flare that illuminated the roofs of the Plant Institute and the other buildings around St. Isaac's Square. Shortly thereafter, the waves of Junker bombers would begin their approach toward the light. Their projectiles churned up the banks of the Neva River, splattering walls and pavements with stains that looked like sprayed blood under the chemical light of the dangling flares.

A new soundtrack accompanied these first days of the siege. In the gaps between radio broadcasts, an engineer at the Art Deco "Radio House" on the corner of Italyanskaya and Malaya Sadovaya Streets would place a metronome before the microphone. The sound of ticking was played through a network of nearly half a million fixed-wire loudspeakers installed in domestic apartments and outdoor public spaces. Whatever suburban territory had been lost to the German advance and however many enemy planes occupied the skies overhead, Leningrad's airwaves remained resolute. The ticking soon provided a steadying rhythm, a pulse for the city, both a confirmation that the radio system was working and, when the mechanical heartbeat quickened, a warning of imminent risk.

"The metronome peacefully taps out seconds," wrote one listener. "It keeps us on alert and anxious all the time any second it could stop, and the siren will sound, and the announcer's voice will proclaim, 'Attention! Attention! Air raid! Air raid!' somewhere over there in the sky which you do not see harvest death the cruelest, basest [death]."

Just as, a few years earlier, the Leningrad-based physiologist Ivan Petrovich Pavlov had trained his famous dogs to salivate at the ticking of a metronome, a sound that had been previously associated in the animal's mind with the sight of food, so the quickening pace of a metronome on the radio prodded the city's residents to find refuge, to push their bodies against a doorframe, or to descend into the thick gloom of a bomb shelter. The procedure became a monotonous routine: the donning of galoshes and overcoats; the mild irritation at having to leave a half cup of tea on the sideboard; the sensation of descent into a chilly cellar along a familiar staircase; the straining to hear, through the thick walls, the muffled chug of the antiaircraft guns.

The moon began to inspire an instinctual response in citizens, too. German air raids became more frequent on clear nights, their bombers better able to aim their explosives by its treacherous light, which exposed the city's vulnerabilities. "How we hated the moon," wrote a second-year student at Leningrad State University who, throughout her later life, suffered insomnia during a full moon when she would lie awake and hear the whistles and thuds.

In the mornings, secure in their childhood sense of immortality, young boys and girls would run through the streets searching for bomb fragments to keep as mementos. Doctors in the city's hospitals had to contend with "shapeless lumps of human flesh, mixed with bits of clothing and brick dust," as Professor Vladimir Garshin, chief pathologist at Erisman Hospital put it, as well as their horrified relatives, shocked into shrieking or silence, who needed official death certificates.

"By evening your soul was paralyzed," Garshin noted. "You were left feeling completely empty."

At the seed bank, a change in tempo on the radio caused the on-duty manager to instinctively open the logbook, ready to record the time of the impending attack. The intensity of the bombing and shelling exerted a severe psychological strain on the Plant Institute's staff, some of whom began to question the wisdom of guarding a collection of seeds when human life was at stake. After one particularly intense raid, one staff member panicked, imploring Ivanov that they should destroy the collection, research papers, and archives to prevent them from falling into enemy hands. "We should mix the seeds together, then burn our papers in the stoker," she begged.

The story of the seed bank's near destruction in the 1920s, when the staff had eaten its nascent collection to save their lives, provided a cautionary fable. It was an understandable but regretful failing that all who served there since had pledged never to repeat.

"The rest of the Institute gave short shrift to such faintheartedness," Ivanov recorded.

III

WHILE COWARDICE WAS DISMISSED AT the Institute, by September it could prove fatal to those charged with the defense of the city. Fearing that Leningrad was about to fall, on September 8, 1941, Stalin had summoned his most talented general, Georgi Zhukov, from the Central Front and ordered him to travel to Leningrad. When he arrived, he was to relieve from his post Kliment Voroshilov, the commander who had failed to prevent the German blockade.

Three days later Zhukov and a small group of military personnel flew with a fighter plane escort from Moscow to Leningrad through gray, rainy skies. As the aircraft broke the clouds and crossed the German lines near Lake Ladoga—the only remaining approach to the city that did not cross enemy lines—a pair of Messerschmitt planes gave chase. Zhukov's pilot dove and, flying low across the water, evaded his pursuers to safely land at Leningrad's military airport.

Forty-four, bald, stocky, and ambitious, Zhukov arrived to find Leningrad's military council in a mood of "drunken defeatism," as he described it. His conference with the city's leaders was brief and evidently convincing. When he left the meeting room, each attendant had vowed to "defend Leningrad to the last man." Zhukov worked quickly, employing drastic measures to strengthen the city's defenses. He ordered the reinforcement and extension of barriers, and the laying of mines across every entry route. At his command the Red Army adapted antiaircraft guns in Soviet-held territory in the suburbs for close-range fire against panzers, not planes. He ordered the transfer of guns from the battleships trapped in the city's docks to the front line.

A week later, around the time that Kameraz fell unconscious in the field, Zhukov signed Order 0064, which decreed that any commander, political officer, or regular soldier caught leaving the line of defense without prior written instruction would be "shot on sight." Likewise, no Russian was allowed to enter the city from enemy-held territory. Stalin ordered troops around Leningrad to fire on any civilians who approached from German lines.

"It is rumored that the German scoundrels advancing on Leningrad have sent forward individuals—old men and women, mothers, and children—from the occupied regions, with requests to our Bolshevik forces that they surrender Leningrad and restore peace, . . . My answer is: No sentimentality," Stalin ordered. "Instead smash the enemy and his accomplices, sick or healthy, in the teeth. . . . Whoever in our ranks permits wavering will be responsible for the fall of Leningrad."

IV

WHENEVER AN INCENDIARY BOMB LANDED on the Institute, four on-duty staff members would race onto the roof. Using a pair of pincers, they would grab the smoking cylinder and fling it down into the courtyard below. One staff member was always stationed in the courtyard, where another staff member was always stationed alongside a pile of sand. The incendiary would hit the asphalt with a cloud of sparks, smoldering while the colleague at ground level rolled it toward the pile to douse it with sand, which would boil from the thousand-degree heat. In total, 108 incendiaries landed on the Institute roof, sometimes causing multiple fires.

All staff who remained in the building were expected to help. Some, such as Professor Georgi Vladimirovich Kovalevsky, who had run high-altitude sowing trials in the mountains of Armenia and south Altai before the invasion, refused to change their dress despite the circumstances. Kovalevsky performed his firefighting duties in a suit, shirt, and tie, maintaining the uniform of his scholarly profession.

When bombs fell in the distance, those standing guard on the rooftop would watch the dreadful spectacle. "Some firebombs fell on the beach by the Peter and Paul Fortress," recalled on observer. "They illuminated the square in front of the Exchange Building while a wooden observation tower on it burned like a torch." But when a raid ended, the seed bank's staff were always eager to return to their research work, an attempt to keep a semblance of normality during these extraordinary times.

"Enough sitting around, time to get back to work" said Shchukin, the groundnuts expert, after the fire team had extinguished an incendiary bomb on the roof and heard the metronome return to its regular tempo. The raids were timed to wear down the population. The Germans would often leave just enough time for people to return home from a shelter before beginning the next bombing raid. The persistent threat of physical assault soon begat mental anxiety. "It is already a form of psychosis," the director of the archives of the Academy of Sciences in Leningrad, Georgi Kniazev, recorded in his diary.

Soon some citizens resigned themselves to the whims of fate. One woman crossing St. Isaac's Square refused to run for cover, despite the cries of concern from people taking shelter in the porchways of the seed bank and other buildings around the square.

"It's all one to me whether I live or die," she shouted in irate resignation. "I'm fed up with everything. It's all hateful."

At that moment a truck carrying a coffin covered in wreaths accelerated past her.

"Now there goes a lucky man," she said.

Around this time, after a raid ended only for the warning siren to herald another, Ivanov grabbed his gas mask and ran down the Plant Institute's staircase into the street. Shchukin hurried along the corridor to join him, in haste struggling to put his arm into the sleeve of his jacket. A deafening crash sounded close by, signaling these were explosive bombs, and not incendiaries intended to spread fire. The Institute's manager, Maria Belyaeva, ordered the staff to take shelter. As Ivanov ran back into the building's entrance, the shock wave from a bomb around the corner slammed him against the wall. He felt the impact of shrapnel smash into the building's stonework close by. As his hearing returned, Ivanov saw a torn gray lump of metal rocking in the street, a fragment of spent shell.

Ivanov was at a loss as to why the Germans had targeted St. Isaac's Square, which had symbolic significance but no obvious military value. Yet a few years later at the Nuremberg trials, Chief of the German Armed Forces Operation Staff *Generaloberst* Alfred Jodl testified to the tribunal that Nazi artillery attacks followed "a carefully worked-out system, according to which only key plants in Leningrad were marked as necessary targets." The heavy artillery was so scarce, Jodl claimed, that "one had to be extremely economical in its use."

A German artilleryman contradicted this testimony. "All the gun crews know that the bombardments of Leningrad were aimed at ruining the town and annihilating its civilian population," he wrote. "[We] therefore regarded with irony the bulletins of the German Supreme Command which spoke of shelling the 'military objectives' of Leningrad."

When the raid was over, Ivanov later recalled, Belyaeva was "beside herself." The roof was pocked with holes, and some of the windows yet to be boarded up had smashed.

"Don't fret," urged Ivanov. "We can replace the glass when the war is over."

"After the war?" answered Belyaeva. "Winter is coming; there'll be snow blowing into the offices, and it's my responsibility."

Ivanov phoned his friend Director Orbeli at the Hermitage to inquire whether the museum still possessed any of the glass the seed bank had provided before the war. Orbeli sent a truckload of offcuts to the Institute, which the staff used to replace the broken windows before boarding them up.

A sense of shared focus and industry reigned among the Institute staff. Then an executive order from Eichfeld clarified the precarity of the staff's situation as well as the chilling absence of sentimentality among those tasked with administrating the Institute's finances: any scientist from the Pushkin laboratory on the outskirts of the city, now behind enemy lines, was to be struck from the register of employees if the person failed to report to the main Institute building by September 16. Captured, killed, or escaped, the employee would no longer receive a salary.

V

"I SUPPOSE THAT SOME PEOPLE are clutching their heads with both hands to find an answer to this question: 'How can the Führer destroy a city like St. Petersburg?'" Hitler told Heinrich Himmler, the leader of the SS, toward the end of September. To rain fire on a city's populace requires the suppression of a person's elemental instincts, a breaking of those fundamental beliefs in the commonality of humankind. Even Hitler relied on twisted logic to overcome whatever scruples he might have felt about launching bombs into a city of civilians.

At first, Leningrad had represented a strategic goal. Its capture would disrupt Soviet industry and logistics, open a direct land route to the northern Soviet Union, secure for Germany the city's arms manufacturers, and facilitate the movement of German troops and resources. Now, Hitler's motivation seemed to have shifted to its symbolic value: razing the city would demonstrate the superiority of the Nazi regime over Bolshevik power and Marxist ideology. "I would prefer not to see anyone suffer, not to do harm to anyone," he continued. "But when I realize that the species is in danger, then in my case sentiment gives way to the coldest reason."

One of his generals summed up this strategic reasoning: leveling Leningrad and Moscow would "relieve us of the necessity of having to feed the populations through the winter" while "depriving not only Bolshevism but also Muscovite nationalism of their wellsprings." German propaganda had long worked to convince its troops of the danger posed to civilization by "Bolshevism," which Hitler portrayed in his screed *Mein Kampf* as a Jewish conspiracy. "It would be an insult if one were to call the features of these—largely Jewish—tormentors of people beasts," read an article in the June 1941 issue of the Wehrmacht's army bulletin. "They are the embodiment of the infernal, of the personified insane hatred of everything that is noble in humanity."

The leaders of the German forces knew they would have to steel their troops for a messier, less principled mode of combat in Russia. In a March 1941 speech to his generals Hitler asserted that the war against Russia "will be such that it cannot be conducted in a knightly fashion." The struggle, he insisted, "is one of ideologies and racial differences and will have to be conducted with unprecedented, unmerciful and unrelenting harshness." Commanders, he concluded, would have to "make the sacrifice of overcoming their personal scruples." When Hitler left the room, some of the officers protested the murderous implications of Hitler's speech, arguing that the planned extermination would violate their soldierly principles and destroy discipline.

By redefining Leningrad's citizens as *Untermenschen*—"subhumans"—moral qualms were less likely to stay the hand of those tasked to drop the bombs. As one of Leeb's panzer commanders, Colonel General Erich Hoepner put it in an operational order to his troops: "This war with Russia is a vital part of the German people's fight for existence . . . and for this reason it must be conducted with unheard-of harshness." Only the complete annihilation of the enemy would preserve the "noble blood" of the German people. The Soviet people were thus dehumanized, the prerequisite to their destruction.

On September 24, 1941, when Germany's forwardmost units were nine miles from the Plant Institute, the German commanders halted their advance. The approach to Leningrad had proven costly to the German military, who had lost around sixty thousand men, but more so to the Soviets, whose dead numbered more than three hundred thousand. The Germans were tired and overextended, but now that they had encircled Leningrad, there was no great rush. A water route via nearby Lake Ladoga stood as the city's only reliable connection to the outside world, an expanse of water across which small planes could fly and, until the water froze, boats could sail. The city had otherwise been encircled by the Germans and their Finnish allies to the north, severed from the Soviet Union and mostly left to fend for itself.

Unsure of what to do, the German Eighteenth Army began canvassing for opinions for dealing with Leningrad's population. Three possibilities emerged, each with advantages and disadvantages: evacuate Leningrad's citizens westward into the German zone of occupation; arrange for their evacuation behind Soviet lines; or, simply, "starve the lot."

The first option would place untenable logistical pressure on the German army to feed and contain millions of Russian civilians. Ensuring the safe passage of Leningrad's civilians into the Soviet Union would have propaganda advantages, but could backfire if, as expected, tens of thousands of Leningraders died en route to Soviet lines. The third option was the most brutal and would provide Germany's enemies with a motivational horror story. And there was no telling what the psychological effect of watching at proximity a city starve would be on rank-and-file German troops. But it was also the most efficient solution, one that would spare the German army the burden of feeding millions. It would be a long wait. Sacking a city takes time; starving a city takes longer.

The choice was soon made for the invaders. On September 29 the chief of the naval staff announced a directive from Nazi high command to his officers:

> The Führer has decided to wipe St. Petersburg from the face of the earth. The existence of this large city will have no further interest after Soviet Russia is destroyed. . . . [Any] offer of surrender will be refused, since the problem of the life of the population and the provisioning of them is a problem that cannot and must not be decided by us. In this war . . . we are not interested in preserving even a part of this city's population.

PART TWO

Leningrad's authorities cut daily bread rations five times between September 2 and November 20, 1941—reducing from twenty-one ounces to nine for workers, engineers, and technicians, eleven ounces to four for adult dependents and children. In all cases this was less than the lowest ration provided in the gulag, where prisoners typically received six times as much bread, as well as soup and buckwheat.

VII

FOUR OUNCES

I

IT TOOK A WHILE FOR the illustrator Grigori Fillipovsky's eyes to adjust to his dim surroundings. The air in cell 27 in Moscow's Butyrki Prison* was dank and filled with the sound of grunts, sniffles, and murmured conversation. As the contours of the space and its occupants shaded into view, the artist estimated that around two hundred men occupied a room built to accommodate twenty-five. These men were not German prisoners of war, but Russian citizens who had been arrested by the NKVD on a variety of treason charges. A single small window provided the only ventilation. The fusty air clung to the men as they jostled for space.

On a dirty bunk in the corner, Fillipovsky spied an elderly looking man lying on his back, his legs raised up. In the gloom the artist noticed the man's face was puffy, with bags under his eyes so severe Fillipovsky wondered if the man was suffering from heart disease. Kindly cellmates had removed the man's boots and placed them beside the bunk, exposing feet that were unnaturally inflamed and deathly pale. Even in this bloated, bedraggled state, the man exerted a certain presence and authority. He was, Fillipovsky guessed, *someone*.

At dawn the following day, guards entered the cell, hoisted the man onto his injured feet, and dragged him away.

---

* Also known as Butyrskaya or Butyrka.

"Who was he?" Fillipovsky asked his cellmates.

"Academician Vavilov," someone replied.

II

VAVILOV HAD SUSPECTED WHAT WAS coming. Nine months before he disappeared in Ukraine, the botanist and explorer had been granted a personal audience with Stalin. Vavilov had had several previous attempts to speak to the Russian leader rejected. He wasn't going to miss this one and arrived in plenty of time for his ten o'clock appointment on the night of November 20, 1939. He waited for two hours in the reception room before, around midnight, he was finally summoned into Stalin's room.

"Good evening, Iosif Vissarionovich," said Vavilov, unaware that the Communist leader disliked being addressed by his first names and preferred "Comrade Stalin."

Stalin ignored the greeting and did not invite Vavilov to sit.

"So, you are the Vavilov who fiddles with flowers, leaves, grafts, and other botanical nonsense, instead of helping agriculture, as is done by Academician Lysenko, Trofim Denisovich."

By contrasting Lysenko's full name and title with the brusque "Vavilov," Stalin clarified his disrespect. Vavilov took a moment to regain his composure, then, while standing awkwardly, began to explain the seed bank project, how he and his colleagues had gathered rare seeds and samples from around the world and brought them to Leningrad to form the basis of a plant-breeding program that, in time, he believed would revolutionize seed production in Russia. As he spoke on the subject to which he had devoted his deepest energies, the botanist felt his anxiety subside. He explained his training of young scientists, and the work being conducted in satellite breeding stations across Russia, where the results of the Leningrad seed experiments could be tested in different types of soil and weather conditions.

Stalin let Vavilov speak for close to an hour without responding. It was, the botanist later told a friend, as if his words were bouncing off a stone wall. Finally, Stalin cut Vavilov off midsentence.

"You are free, Mr. Vavilov," he said, waving his guest away.

Having clung to the hope that, if he could just speak to Stalin, he could convince him of the value of his work, Vavilov's optimism vanished. When he returned to Leningrad, he told a friend that there was no hope left for the genuine pursuit of plant science.

"We must hurry to produce immortal works," Vavilov wrote to his colleague Professor Konstantin Pangalo, whom he had encouraged to come back to work after the professor lost his leg in a tram accident and fell into a deep depression. Time, Vavilov knew, was short. Still, his arrest came as an aging force, a shock that knocked the present into the past and, as the Russian novelist Aleksandr Solzhenitsyn put it when describing his own, similarly anticipated yet unbelievable arrest, ushered "the impossible into omnipotent actuality."

III

Six days after NKVD agents bundled Vavilov into a car outside the youth hostel in Ukraine, Aleksei Khvat, a thirty-three-year-old senior lieutenant in the Soviet secret police, read out the charges leveled by the state against the director of the Plant Institute.

To those who knew, loved, and followed Vavilov, he was a scientist who combined principled research—the willingness to allow experiments to disprove his theories, in the understanding this would help him improve them—with personable qualities of charm and genuine care for his understudies. Lieutenant Khvat's task, as he understood it, was to dismantle this charade and expose Vavilov as a duplicitous spy who worked for the British.

"You have been arrested as an active participant in a subversive anti-Soviet organization and a spy for foreign intelligence services," Lieutenant Khvat said, on the morning of August 12, 1940, in the first of several dozen brutal interrogations. "Do you admit your guilt?"

In Stalin's Russia the arrested individual was guilty until proven guilty. The surest way to earn a death sentence was to deny the state a confession. Thousands of men who rejected the administration's fiction perished under the rigors of preliminary investigation or were shot after a brisk and closed trial. A person's best chance of survival was, counterintuitively, to admit to everything. Even then, admission was rarely sufficient to save one's life.

Even knowing all of this, Vavilov maintained his innocence in the firmest terms. Yes, he had traveled internationally. Yes, he had established links with farmers and dignitaries across the world. But this work, he insisted, was only concerned with the spheres of botany and science, not international politics, or war.

"I declare categorically that I have never engaged in espionage or in any other kind of anti-Soviet activity," he stated. Lieutenant Khvat read out wounding testimonies from Vavilov's colleagues in Soviet science, accusing him of "wrecking activity in agriculture." These were, Vavilov insisted, "one-sided," "incorrect," and "slanderous."

The brutality of the Soviet state in seeking confessions from those it deemed to be "spies and wreckers" had been honed and sharpened through years of practice. Interrogators would force elderly men to lie facedown, then beat the soles of their feet and spine with a rubber strap, creating bruises that would, in later days, provide targets for further violence. Inquisitors devised arcane modes of torture. In one peculiarly cruel and baroque method, known as the swan dive, the interrogator would insert a long piece of rough toweling between the prisoner's jaws like a bridle. The torturer would then pull the ends back over the prisoner's shoulders and tie the material to his heels so that he lay on his stomach like a wheel, spine pulled taut against its natural curvature. The victim was then deprived of food and water for two days.

If a prisoner proved especially resistant, Soviet interrogators would often produce a stranger whom they threatened to kill in the absence of a confession. The interrogators worked in shifts, depriving their victims of sleep. They would place bowls of hot food in front of the starving; some prisoners claimed that their interrogators had urinated in their mouths. This commitment to their cruel work bespoke the pressure these men felt from above; Stalin reportedly followed the progression of certain interrogations for years. He claimed to be unable to sleep until certain confessions had been secured.

The point was not to extract truth from liars, but to force inno-
cent people to collude in a fiction. Still, Vavilov's resolve held, even as
Lieutenant Khvat's interrogations lengthened to ten, twelve, thirteen
uninterrupted hours. Guards would fetch Vavilov in the early hours
of the morning, when he was dozily vulnerable, then question him
until daybreak. As deputy head of the investigation department of
the NKVD's economic section, Lieutenant Khvat knew that his future
career, maybe even his own life, depended on Vavilov's profession of
guilt.

As the questionings lengthened, Lieutenant Khvat's written minutes became shorter, cloaking the physical and psychological violence
of each meeting in the sterile sheen of administrative record.

"Which countries have you visited?" Lieutenant Khvat asked in
one session, the only recorded question in an interrogation that lasted
ten and a half hours.

Either through fatigue, humiliation, injury, sleeplessness, or
threats toward his family and colleagues, on August 24, 1940, eighteen
days after his abduction and arrest, Vavilov's resolve broke. He offered
his first false confession, perhaps the only way to now spare himself
from death by torture. He claimed that, since 1930, he had belonged
to a right-wing, anti-Soviet cell working inside the People's Commissariat for Agriculture. Vavilov, who had spent his years mounting expeditions to collect and preserve vulnerable plant specimens from the
farthest reaches of the world had, he now lied, spent the past decade
plotting the breakdown of Soviet agricultural science.

When pressed as to who else had been involved in these schemes,
Vavilov was careful to name only acquaintances who had already been
executed as enemies of the people; even in his besieged state, he sought
to minimize injury. Finally, and most self-woundingly, he claimed that
he had established and run scientific institutes such as the Plant Institute with a view to "harm agriculture." It was due to his treachery,
he lied, that the Soviet Union had experienced the famines of 1931, a
scandalous admission in every sense. The officer recorded each deli-
cious confession in shorthand, then wrote them up as statements for

his prisoner to sign. Vavilov stopped short, however, of admitting to having passed state secrets to foreign powers.

"I am not guilty of espionage activity," the confession concluded.

Lieutenant Khvat continued his daily regimen of arduous interrogation.

Next, on September 11, 1940, Vavilov confessed to being a "wrecker." He signed a document with a title that left no room for doubt: "Wrecking in the System of the Institute of Plant Breeding That I Directed from 1920 until My Arrest on the 6th of August 1940."

The botanist's purported admissions were obvious falsities. The suggestion that he had "wrecked" an institution that he had, in fact, built up from a leaderless, furniture-less building where much of the seed collection had been eaten was indefensible. Policies he claimed sole responsibility for were included in plans and resolutions that had been approved at Party meetings. Even a cursory fact check of his statements would have shown that Vavilov was attempting to accept blame and responsibility for government-led policies and deceptions that had led to the breakdown of his own experimental work.

Lieutenant Khvat was uninterested, however, in Vavilov's true role in events. His false admissions were now combined with the denunciations by his enemies and added to dossier no. 268615, a file of evidence against Vavilov first opened in 1931, and which had by now grown to seven volumes.

Some of these denunciations were later shown to be forgeries; others were provided by colleagues who had been recruited by the NKVD as informants in return for pardons or favors. Together with Vavilov's signed confessions, drawn from a total of nine hundred hours of questioning, the document provided sufficient evidence to meet the expectations placed on Lieutenant Khvat, who now indiscriminately burned Vavilov's papers, including ninety notebooks in which he had recorded in detail his expeditions and drawn maps of plant distribution—an act of appalling and willfully misdirected vandalism.

Having delivered his confession, Vavilov was left in his cell to rest. For the next seven months he wrote the first draft of a new book titled "A History of the Development of Agriculture," using the paper and pencil the NKVD had left in his room for any further admissions. Then, in March 1941, the interrogations began again. In the intervening months Lieutenant Khvat had arrested five of Vavilov's alleged accomplices, at least two of whom Vavilov barely knew. Exposing a cell of conspirators was, it seemed, of greater value to the authorities than acquiring the confession of a single agent. This second round of questioning was more physically violent and psychologically humiliating than the first.

"Who are you?" Lieutenant Khvat would ask at the beginning of each session.

"I am Academician Vavilov."

"No. You are a piece of shit."

IV

BY THE TIME OF HIS arrest Vavilov had two families. He met his wife, Katya, when they were both students; she had been the cleverest and best-educated student at Petrovsky Agricultural Institute. They bonded over a shared talent for science, married, and, in 1918, had a son together, whom they named Oleg. Within two years the marriage, which was built more on intellectual kinship than romantic interest and intimacy, had broken down. When Vavilov arrived at the Plant Institute in 1921, he had already begun to correspond with a former student. Yelena Barulina had been one of Vavilov's students when he first came to the city as a pioneering biologist. A twenty-two-year-old orphan with dark hair and deep eyes, Yelena was in her third year of studies when the pair fell in love, just as his marriage to Katya fell apart. He and Yelena did not marry, but did move in together. In 1928 Lena, as Vavilov affectionately named his partner, gave birth to his second son, Yuri.

Shortly after Vavilov's arrest, NKVD agents visited Lena and her son, Yuri, who was by now twelve years old. At the time of the visit they were staying at the experimental farm in Pushkin. As the German troops advanced on Leningrad, the pair fled inland, eventually arriving at Saratov, where Lena had grown up. Neither of Vavilov's families—

Katya and Oleg in Moscow, and Lena and Yuri in Saratov—knew of his whereabouts. Both families now found themselves in an unenviable position. Close relations of any individual charged with treason were cast out of society, unable to secure jobs or state benefits. Vavilov's fate was not only his own; it would dictate the fortunes and futures of those he most loved.

On July 9, 1941, two days after staff at the Plant Institute were ordered to help build Leningrad's fortifications, Vavilov stood trial. Among the pile of evidence that Lieutenant Khvat had gathered, extorted, and fabricated was the signed opinion of five leading experts who together denounced their colleague's work. All five experts were, it later transpired, informants or supporters of Lysenko, who had personally chosen the panel's members. The expectation was clear to all. They were tasked, as one member later admitted, to "provide a deliberately prejudiced and negative opinion of Vavilov's work."

At court, three generals passed sentence on Vavilov without calling on a single witness, any testimony or piece of evidence, or lawyer. The trial lasted only a few minutes. The sentence had already been drawn up. The chairman pulled the document from his briefcase and read the statement aloud. It contained a litany of false charges against Vavilov, including treason. The document concluded with a statement of grim finality:

> The Military Collegium of the Supreme Court of the USSR sentences Vavilov, Nikolai Ivanovich, to suffer the supreme penalty—to be shot and all his personal property to be confiscated. The sentence is final and not open to appeal.

Lieutenant Khvat had discharged his duties with diligence, energy, and clarity of focus that, had he earnestly been seeking the truth, would have been commendable. Despite the chairman's ruling, there was one last route of appeal open to Vavilov: the Bureau of the Supreme Soviet Legislature. In his cell that evening a guard presented the botanist with a pen, ink, and the relevant form.

He wrote:

> After devoting thirty years of scientific work to plant breeding, I ask that
> I be given some small chance to complete my work for the benefit of
> socialist agriculture in my motherland.
>
> As an experienced teacher, I undertake to totally devote myself to the
> training of Soviet specialists. I am fifty-three.
>
> 20.00
>
> 9.VII. 1941.

Seventeen days later the Presidium of the Supreme Soviet rejected Vavilov's appeal, the 283rd petition heard that day. Guards transferred the botanist to Butyrki, the largest of Moscow's three main prisons, set aside for "politicals," an institution filled with academics, scientists, and novelists.

Two of Vavilov's close colleagues from the Plant Institute were immediately executed for their support: Georgi Karpechenko, the first person in the world to produce a fertile cabbage-radish hybrid, and Leonid Govorov, a leading pea expert who, upon learning of his closest friend Vavilov's arrest, traveled to Moscow to seek an audience with Stalin. Beaten, swollen, and hopeless, Vavilov lay in his cell and awaited news of the date scheduled for his killing, while the illustrator Grigori Fillipovsky and his other cellmates watched on.

V

ON SEPTEMBER 8, 1941, JUST as the last road out of Leningrad was cut and German bombs destroyed the Badayev warehouse and filled the air with the scent of bacon and sugar, a junior trade commissar, Dmitri Pavlov, arrived by plane, flying across Lake Ladoga from the east—the only way left for a person to reach the encircled city. He began to compile an inventory of the food stocks held in warehouses, factories, and other storerooms and depositories. The results of his calculations were foreboding. At the current rate of consumption Leningrad's food reserves would last for little more than a month.

There was enough flour to feed the population for thirty-five days; meat would be gone within thirty-three. Rice, semolina, and pasta would run out within thirty. Only sugar and confectionary were to be found in relative abundance: sixty days' worth—not quite enough to see them through to Christmas. The German domination of the skies over Leningrad made plans to deliver food via the Lake Ladoga approaches dangerous; Stalin refused to authorize an airlift of food.

The failure of the authorities to adequately prepare for a blockade had left residents woefully vulnerable. None, seemingly, had anticipated a siege, and Stalin had expressly instructed staff responsible for trade and supply not to divert extra provisions to Leningrad beyond previously agreed quantities.

"How was it that they didn't have more foresight?" mourned one observer. Unless the siege broke soon, Pavlov surmised, the people would be bereft of resources.

Having failed to stockpile provisions for a blockade, Leningrad's leadership now raced to gather food from farms situated within the siege ring. Farmworkers were allowed to keep thirty pounds of potatoes a month but were compelled to surrender the remainder to requisitioning parties, which included conscripted Leningraders sent out to collect grain and vegetables. These parties toured farmsteads like roaming gangs. Any peasant found hiding potatoes was subject to the same wartime punishment meted out to deserters, while private stores of grain were plundered via "compulsory donations."

In mid-September the safer roads into Leningrad were filled with hundreds of citizens carrying sacks and baskets of farm produce. These supplies were supposed to enter the rationing economy, but officials feared that much was "squandered into private hands."

Attacking millions of Russians with the blunt weapon of starvation had been part of the German strategy from the start, a goal made explicit in its name: *Der Hungerplan*. Devised by the zealous Nazi and State Secretary in the Food and Agriculture Ministry Herbert Backe, the Hunger Plan aimed to solve the considerable logistical problem of how to feed an invading army of four million men and six hundred thousand horses.

Germany, like Britain, is a nation reliant on imports to feed its people. The memory of the First World War, when more than half a million Germans, mainly women and children, starved to death following a British blockade on imports remained fresh and rhetorically persuasive. No home front can survive without food. The invasion of France had disrupted the production and distribution of foodstuffs on which the German population was reliant, as had Britain's naval blockade. Operation Barbarossa threatened to exacerbate food shortages that were already affecting the Reich's civilian populations.

To divert provisions from Germany to feed millions of soldiers trekking across the Soviet Union was impractical. So Backe and his senior officials planned for the troops to take food from the hands of local populations to sustain the invaders. When Ukraine fell, provisions could be redirected from the Soviet Union's industrial heartlands to the German army, thereby easing hardship at home. Ukraine's grain surplus, while modest, currently fed millions of residents in Russia's cities. Remove these cities from the supply chain and, Backe reasoned, that food could meet Germany's ever-expanding needs.

It was a cruel plan, but in character for Backe, who had already demonstrated his willingness to use famine against civilian populations during the invasion of Poland. The cost to civilian life was clear to the German command. This weaponized famine, according to Nazi documents collated in a policy file that would later become known as the Green Folder, would devastate the lives of ordinary Russians:

> Many tens of millions of people in this territory will die or must emigrate to Siberia. Attempts to rescue the population from death by starvation by drawing on the surplus from the Black Earth regions can only be at the expense of the food supply in Europe. They diminish the staying power of Germany in the war and the resistance of Germany in Europe to the blockade. There must be absolute clarity about this.

There would be suffering. German leaders were given explicit instructions to hold firm when faced with scenes of ravenous agony.

Anyone moved to distribute a portion of their food rations to starving Russians would threaten the survival of the Reich and the success of the war. Loyalty required a hardening of heart. German soldiers would preside over the starvation of Russian civilians as a matter of national priority. In this way, the Hunger Plan, which was approved by the heads of the Reich's key ministries a few weeks before the invasion, provided the blueprint for genocide.

Now, as German troops forged into Soviet territory, the front line systematically severed millions of Russians from their primary sources of food, leaving hunger in its wake. And the seeds of famine were sown with the encirclement of Leningrad, where three million civilians, soldiers, and sailors were trapped in the city and its surrounding villages.

To compound issues, jurisdiction over food supplies was divided between ten different agencies, which continued to work independently of one another in their distribution of provisions. Scarcity immediately began to heighten inequality.

In the summer, wealthy Leningraders had been able to stockpile foodstuffs by purchasing supplementary items from the expensive "commission shops," set up to reassure citizens with the sight of well-stocked shelves. By September most of these stores had closed, but a gray market of off-ration supplies was readily available in many canteens and restaurants. Wealthy individuals could purchase oil, butter, meat, and sugar, as well as luxury items such as caviar, coffee, and champagne, under the counter. Fifteen commercial restaurants continued to sell food in unrestricted amounts until mid-November, diverting essential resources away from the rationing system toward the cupboards of the wealthy.

For the ordinary citizen, the effects of rationing were swift and keenly felt. The authorities cut daily bread rations five times between September 2 and November 20—reducing from twenty-one ounces to nine for workers, engineers, and technicians, eleven ounces to four for adult dependents and children. In all cases this was less than the lowest ration provided in the gulag, where prisoners typically received twenty-five ounces of bread plus soup and buckwheat. Even those prisoners subject to "punitive" measures received eleven ounces.

To compensate for these reductions, rations of fat and sugar were increased in September, a policy that stored up suffering and woe for the winter, when the two and a half tons of sugar and six hundred tons of fat could have helped save tens of thousands of lives. New recipes attempted to turn basic ingredients into palatable dishes: shells of sunflower seeds were ground to make soup; ersatz coffee grounds were shaped into pancakes known as *gushchaik*. "They are extremely unappetizing," one resident recorded in her diary. "And after you eat them: heartburn."

Like bankruptcy, starvation crept on the city slowly at first, then swiftly and decisively. When, on November 8, 1941, the tiny town of Tikhvin, situated on the eastern shore of Lake Ladoga, fell to the Germans, another key strategic hub disappeared. It was through Tikhvin that freight had, until then, dribbled to the wharves and jetties on the eastern shoreline of the lake, to be shipped across the water into the city. The choking of this minor but vital route squeezed Leningrad further. The battlefield was close enough that some soldiers would attempt to return to the city at night to share their rations with their starving families.

The day after Tikhvin fell, Hitler delivered a radio broadcast in which he made his plan for Leningrad explicit: "The enemy will die of starvation."

VI

EVERY NIGHT ABRAM KAMERAZ AND his pregnant wife, Lyudmila, guarded the metric tonne of rare potatoes he had rescued from the Pavlovsk field station. These specimens were excluded from the city's rations for good reason: Kameraz and his colleagues believed that, when the siege broke, the tubers could be used to replant the fields and begin propagating a harvest that could in time feed not only the people of Leningrad, but people everywhere.

A cold, dank cellar with hard floors and only burlap sacks for blankets was an uncomfortable arrangement for any person, but especially a pregnant woman. Still, Kameraz grew fond of his new living quarters, which offered a kind of dark, quiet sanctuary amid the gathering storms of conflict.

"It was," he later recalled, "tranquil there."

Then, on November 2, 1941, six weeks after he had fetched the last of the haul of tubers, Kameraz received a summons to the front. Fit and able-bodied, he would serve as a mortarman in the 123rd Rifle Division, a member of the Twenty-Third Army. As a multilinguist who spoke fluent German, Kameraz would moonlight as an interpreter and, perhaps learning of his daring escape from the field station in Pavlovsk, his superiors soon assigned him to reconnaissance missions. He bade goodbye to Lyudmila, not knowing if or when he would see her again or have the chance to meet and hold their child.

Around this time, his colleagues received an order to box up a small part of the potato collection to be loaded onto a small freight plane and removed from the city. The train laden with seeds remained on the sidings, unable to move in either direction; but a small plane might evade the German lookouts and escape across Lake Ladoga. Kameraz, unable to accompany the cargo for which he had risked injury and death, passed responsibility for its safeguarding to his friend and colleague S. M. Bukasov, doctor of agricultural and biological sciences, one of Vavilov's colleagues who had first brought the specimens to Russia from South America.

In Russia, November 7 was a significant date: the anniversary of the Bolshevik Revolution and a major Soviet public holiday when families celebrated with feasting and drinking. With empty warehouses and cupboards, in 1941 Leningrad there would be few banquets, but the German army planned to intensify its attacks as a symbolic gesture. For days in advance enemy planes dropped leaflets warning residents, "Go to the baths. Put on your white dresses. Eat the funeral dishes. Lie down in your coffins and prepare for death. On November 7 the skies will be blue—blue with the explosion of German bombs."

While the German troops prepared their bombers for a night of attacks, and the Leningrad command attempted to conduct preemptive targeted attacks on nearby Nazi airfields, the botanists prepared to evacuate the potatoes before the bombing intensified.

On the night of November 5, 1941, as the population increasingly felt pangs of hunger, Bukasov clambered into the plane and set off for Krasnoufimsk in the Ural Mountains. There was only room for a portion of the collection to be evacuated; the remainder would stay in the cellar. The plane crossed Lake Ladoga under cover of night, evading detection by German fighters. Two nights later the Germans dropped enormous flares on parachutes to light up the parade-less city, as their bombers began to drop naval magnetic mines, each weighing a metric tonne or more, attached to parachutes.

VII

WITH KAMERAZ GONE, RESPONSIBILITY FOR guarding the potato collection now fell to Olga Voskresenskaya, who had helped rescue potato samples and scientific equipment from Pavlovsk. Voskresenskaya was one of the thousands of women who assumed the roles and responsibilities of the able-bodied men under the age of fifty-five who were summoned to the front. As well as performing rooftop watches during air raids, known as *dezhurstvo*, extinguishing fires, and helping to pull bodies from the rubble, women volunteered or were assigned to take men's places in factories, joined coal workers' battalions, and provided primary medical care to the wounded and dying.

Voskresenskaya had joined Vavilov's staff in 1930, when she was just twenty-six. Despite an adverse childhood spent in an orphanage, she was bright and tenacious. She obtained her PhD at the seed bank, worked her way up from lab technician to her current position of senior research officer, a specialist in botanical taxonomy. Whenever Vavilov and his expeditionary team brought new varieties of potato back with them, the specimens were placed into quarantine to prevent the spread of diseases. Vavilov entrusted Voskresenskaya with conducting regular quarantine inspections.

Now, her duty shifted from examiner to protector. As food became ever scarcer, the potatoes attracted attention. Soon, mortal hunger would lead desperate people to loot neighbors' houses, tear the bread from the hands of weaker citizens, and, as desperation turned to savagery, even provoke murder to steal the victim's ration card. Anticipating these effects, Voskresenskaya began to keep watch alongside two junior staff members, lab technicians Klavdiya Chernyanskaya* and A. P. Kovalenko.

Each day and night the three women protected the cellar from rats and thieves, while praying that the Red Army would break the blockade and push the German troops back from their artillery positions. Not only were the staff becoming weaker, but the tubers were in danger from the coming winter frosts; if they failed to survive, then Voskresenskaya's and Kameraz's work would all be for nothing.

The Hermitage Museum across the street from the seed bank was also guarded by women employees. "My mighty army consisted mostly of women of retirement age, including those in their seventies," recalled the museum's head of security. "The guard consisted of around thirty women. Those were my troops, indeed." For many women, their duties were assumed alongside the responsibilities of keeping a house and raising children.

Evgenia Kuznetsova was charged with guarding the wheat samples at the Plant Institute. She had said goodbye to her husband and son a few weeks earlier, when the family evacuated from the Pushkin field station. After the harvest was gathered that summer and the Germans advanced on the city, Kuznetsova and her daughter had traveled to Leningrad to live in the seed bank. She did not know it yet, but she was already twice bereaved. Her husband, Fyodor Nikitovich Prokofyev—a research officer at the Pushkin laboratory, who had installed a flour mill there to test the yield of different varieties of flour—had been killed as a member of the home guard. Her son, Askold, a keen hunter, had enlisted as a sniper. On September 12, the day Pushkin fell to the advancing German army, he sent his one and only postcard home.

* Spelled in some other primary source documents as K. T. Cheryavskaya.

Kuznetsova had worked with Vavilov since her graduation from Saratov University in 1923, leading experiments to see which varieties of wheat were best suited to different Russian climates. She was used to conducting scientific experiments in environments dominated by women; most of her classmates at Saratov University had been young women, as the men had all been called up to fight in the Civil War. Now, once again, she conducted her work alongside female scientists such as Voskresenskaya, overseeing the junior female researchers Serafina Arsenievna Shchavinskaya and Lydia Mikhailovna Rodina, who had been given responsibility for the oat collection.

The urge to continue working to maintain a semblance of normality was common to both men and women. Director Iosif Orbeli ordered that the Hermitage remain open to visitors, even after its most valuable treasures had long left the building. Whenever soldiers came to the museum to clean up fallen debris after an artillery strike, or to replace windows with plywood, or to retrieve bodies, Pavel Gubchevsky would invite the men to join him on a tour of the building. He would lead the soldiers from room to room, pointing at the empty frames and forsaken plinths, while describing in vivid detail the piece that usually hung or sat in place. "His descriptions were so vivid," one witness recorded, "that they could almost see Rembrandt's *Prodigal Son* and da Vinci's *Madonna*."

The academic and archaeologist Boris Borisovich Piotrovsky worked on a thesis while, outside the museum, the bombs and shells fell. On October 23, 1941, Piotrovsky was attending a concert of Tchaikovsky's music when a heavy artillery bombardment began.

"I was fascinated by the chandeliers shaking from the explosions while the orchestra was performing the *Capriccio italien*," he later recalled.

Work at the Plant Institute continued apace, too. In October 1941, two more orders came through, instructing the staff to produce inventories of the assets rescued from both the Pushkin and the Red Ploughman facilities, and the property and assets of the Institute. Georgi Nikolaevich Reuter, who had worked as head of the human resources department at the Institute since the summer of 1939, was responsible for the safety of the seed collections and equipment and oversaw the inventory taking.

In the building's chemical laboratory, staff developed methods for extracting pharmaceutical tannin from the leaves of the *Cotinus*—smoke bush—to be used in the treatment of wounds in military hospitals. Five Institute employees who had been requisitioned to work at one of the Leningrad factories oversaw the production of medical tannin—the first time it had been produced at scale in Russia—to be sent to the front. And a special assignment arrived from the military requesting that Institute staff develop methods for using local plants to dye fabric khaki, to be used by soldiers on active duty. Experts in the agrometeorology department produced reports on the best ways to transport troops and supplies under various weather and soil conditions. And as the threat of starvation loomed, scientists explored how nutrients might be extracted from food industry waste products to extend the city's rapidly depleting resources.

As the temperatures outside began to fall and fuel became increasingly scarce, it became harder to heat the Plant Institute's warren of rooms and corridors. During the early autumn months, staff had removed panes of glass from the windows to reduce the risk of injury during the German bombing raids. The replacement wooden boards were safer, but did not keep out the wind, which began to whistle cold through the corridors. The lack of carpets and ineffective insulation meant there was barely a difference in temperature inside the building and out. As the remaining staff grew hungrier, the cold heightened the difficulty the botanists felt in focusing on their work.

Nikolai Ivanov felt reduced. He was unable to concentrate, suffered intermittent blackouts, and would often need to lean against the wall to rest whenever walking between the Institute's rooms. He developed a permanent stoop, his body's instinctive attempt to ease the constant ache in his stomach. The bags in the hollows beneath his eyes turned to bruises.

As a child Ivanov had experienced the disruptive effects of collective hunger. In the summer of 1918, having finished sixth grade at the Fourth Larinsky Grammar School, he and his brother were sent away from Leningrad as part of the "children's food colony" when several hundred children were evacuated from the starving city. A three-month break turned into a two-year absence, as the children were cut off from their families by the White Guard and its allies in the Civil War, and were taken, first to Japan, then North America, before finally returning to Russia. Now, once again, famine had come to his home.

Understanding that the survival of the Institute's staff was contingent on their support for one another, Ivanov made regular trips to the basement where Voskresenskaya and her colleagues had established their guard post.

"It's a long walk from my flat," she told Ivanov. "And I save on fuel. I'd have had to burn it at home, whereas the specimens here in the basement need the warmth more."

Deprivation had already taken its toll. Voskresenskaya's face was aged by malnourishment and exhaustion, hollowed out by the cold winds of material poverty. To Ivanov she seemed translucently pale, like a child pulled from the sea.

She boiled a kettle of water on the stove as the pair chatted.

"Oh, well," she said. "Let's have a glass of hot water." Hearing her convulsive cough, Ivanov urged his friend to rest in the infirmary.

"Who would stay here?" she protested. "What if there's a hard frost? The precious collection will be lost. I don't need to explain this to you, do I?"

As he left, Ivanov kissed his colleague's frozen hands. He walked across the square to the Institute. When he reached his office, he opened a cupboard where he kept some rawhide harnesses. Now, in the grip of famine, the leather had proven useful in an unexpected way. When cut into tagliatelle-like strips and boiled for eight hours or so, the material would soften sufficiently to become edible.

Ivanov cut off about a yard's worth of leather, wrapped it in an old newspaper, and thrust it into his coat pocket. He returned to Voskresenskaya's cellar. Unsure of how to present her with the gift in such a way she would not refuse it, he passed the bundle to one of the lab technicians with an instruction to cook it for the team. After weeks of eating little else but corn seed husks, the meager meal seemed, Voskresenskaya later recalled, like a king's feast.

VIII

IN THE CONTEXT OF ALL-CONSUMING need, the smallest act of kindness was elevated to greatness. And yet, Ivanov's leather harnesses could only provide modest relief. Shared rations were insufficient to stem the incoming tide of suffering.

Grief came early to the seed bank. In early November staff learned that the seed bank's maintenance manager, Grigory Golenishchev, had been killed by a German shell. He had been an active member in the building's self-defense group, extinguishing fires started on the roof by incendiary bombs. But there was little time to grieve Golenishchev's death before catastrophe again struck.

On November 11, 1941, Pavel Pavlovich Gusev, an adviser at the Institute and Vavilov's former personal secretary, collapsed and became the first member of the seed bank's community to die of starvation. On the same day, just two months into the siege, Maria Parfinievna Dmitricheva, a bibliographer and one of the Institute's librarians, who had been appointed deputy superintendent of a bomb shelter close to the seed bank, also died of emaciation; her taut skin revealed her ribs, as neatly fragile as those of the carcass of a tiny bird. The colleagues had died while their place of work housed a bounty of seeds that could have saved them.

A pitiful irony: the plant-breeding programs nurtured and overseen by Vavilov had been instituted to improve agricultural production. Through selection of preferred plant traits such as resistance to certain pathogens and improved protein content, the seed bank project sought to provide insurance against starvation. But denied access to the fields required to plant and harvest, the seed bank itself faced destruction from the same forces it sought to confound. Through the fug of cold and exhaustion, the mortal stakes of the decision Vavilov's staff had collectively taken to protect the collection for some future goal, rather than to prolong the actual lives of the people around them, snapped into harrowing focus.

The collection represented not just the possibility of new life but a precious link to the past and the history of the land. To distribute these specimens would be to sacrifice this connection, to foreclose the possibility of ever reclaiming what had been lost. And yet, death had come to Leningrad. To withhold the seeds would betray the same people the botanists sought to serve.

A harsh and terrible winter loomed.

In the terrible winter of 1941–42, when temperatures fell as low as -40°C (-40°F), Leningrad's residents burned books, furniture, parquet flooring, and even dried-out feces for heat. Staff heated the cellar at the seed bank to ensure the potato specimens stored there did not perish the frost, using wood pulled on children's sleds.

VIII

CITY OF ICE

I

V<small>ADIM</small> L<small>EKHNOVICH</small> <small>OPENED HIS FRONT</small> door and stepped into an unimaginable cold. The botanist had experienced dozens of Lenin-
grad winters, but few so brutal as this one. Temperatures had begun
to drop below -40°C (-40°F), at which point the hairs in your nos-
trils freeze and your breath turns to a cloud of crystals that tinkles
to the ground with a sound Siberians call "the whispering of the
stars."

En route to the seed bank Lekhnovich walked close to the buildings,
taking narrow side streets wherever possible, to reduce his time in the
open. He carried in his pocket a crude map that marked the estimated
positions of the German artillery. The streets were quiet, interrupted
only by the odd cadaverous shape that shuffled in the flintlike cold.
Blameless snow creaked underfoot. Still, it seemed the weather itself
had joined Leningrad's enemies in their assault. The army was at the
gates, but in no hurry to invade a place that with each passing day
became weaker and more desperate.

Months before the invasion, the traumatic disappearance of
Lekhnovich's friend and mentor Vavilov, combined with the increasingly strained political atmosphere at the Plant Institute, had been
too much for him to bear. When he returned from the fateful seed-
collecting mission in Ukraine, he had resigned his position and, in
January 1941, four months after Vavilov's abduction, assumed the role

of editor of the scientific journal *Priroda*, working out of the Academy of Sciences publishing house.

The invasion had a focusing effect on the thirty-nine-year-old, who had spent his childhood in the village of Dolzhok, close to the front during the First World War. In 1921, when millions went without food, he had joined the *prodarmiya*, a quasi-military group entrusted with procuring bread and fuel for the famine-stricken *povolzhy*—those who lived along the banks of the Neva. Most of Lekhnovich's formative years had been lived against a backdrop of deprivation and conflict. He had seen starvation's dreadful effects on the human body, but still, the sights and sounds of war were familiar; he knew how to survive here at the extremities of human experience.

As editor of *Priroda*, Lekhnovich had turned his editorial focus to subjects relevant to any reader living inside a besieged city under daily threat of bombardment. He published articles on the effect of explosions on the human ear, how to ensure the proper ventilation of bomb shelters, the risk posed by shattered windowpanes during bomb blasts, and, as food supplies dwindled, how to maintain health amid nutritional deficiency. Even though he had resigned from his position at the Plant Institute, Lekhnovich held deep emotional ties to the place. It was in these buildings he had pioneered the study of cancer-resistant potatoes with his close friends Kameraz and Buka-sov. And it was in the seed bank's dim and storied corridors that he had met and fallen in love with his formidable wife, Olga Voskresen-skaya.

When, in early December, Voskresenskaya caught a cold, its effects were heightened by the combination of her lean frame and se-vere malnourishment. She abandoned the cellar where she had been sleeping alongside the potato collection and retreated home to bed. With Kameraz fighting at the front, and his wife recuperating at home, Lekhnovich made the difficult decision to return to the seed bank, and all the complicated memories it held. He would oversee the preserva-tion of the potato collection that his friend and his lover had together battled to save.

With the arrival of winter, the cellar had become too cold for human habitation. To live among the potatoes now could prove fatal. So each morning Lekhnovich bade goodbye to Voskresenskaya and struck out from the freezing house on Nekrasov Street for St. Isaac's Square, two and a half miles away. He would return to care for his wife at lunchtime, before returning to the seed bank a second time in the afternoon. The combination of extreme cold and extreme hunger made his mission grueling. "It was difficult to walk," he later recorded. "It was unbearably difficult to get up every morning, move your hands and feet." Yet, driven by a sense of duty and responsibility to his wife, to his colleagues, to his vanished mentor, each day Lekhnovich faithfully set out on his long, difficult commute to the cellar.

Lekhnovich knew a great deal about the characteristics of the potatoes he guarded. He had collected many of the specimens himself during field trips to the farthest reaches of the world. His PhD dissertation, published in 1937, was titled "Chilean and Andean Potatoes and Their Use in Plant Selection." Lekhnovich understood that the potatoes had to be planted and cultivated if they were to survive the cold season. He planted the tubers in soil in the basement and sealed the doors with three different locks and a heavy iron wedge as a security measure. Neither man nor animal could enter the basement without Lekhnovich's help.

He also faced a different kind of enemy, a spectral attacker unbound by solid locked doors or shards of glass. If the temperature in the cellar fell below 2°C (36°F), the potatoes would succumb to frost and die, and all his wife's and Kameraz's efforts during the past six months would have come to nothing. Lekhnovich needed a way to warm the stones beneath the Institute.

II

BY THE TIME LEKHNOVICH ASSUMED responsibility for the rescued potatoes, the battlefield had begun to extend to his body. The loss of food stocks burned in the attacks on the Badayev food warehouses in early September had been compounded by enemy attacks on barges attempting to cross Lake Ladoga in the weeks before the water froze. Divers salvaged some of the sprouted grain from the water, which could be dried and remilled and mixed with good flour, but almost no food made it safely to the city via water. From October, nonworkers were supposed to receive one-third of a loaf of bread per day. Many received a much smaller ration. Inexplicably, Stalin refused to authorize an airlift of food until mid-November.

Hunger now became the defining sensation in the city, an obsession that superseded all other thought and consideration. Obliged to live on dwindling portions of bread and a five-and-a-quarter-pound ration of food per month—the equivalent to a two snack-size Mars bars' worth of food—households began to use up their cupboard supplies of rice and pasta. Those who had survived the periods of famine during the Revolution and then in the thirties, knew to divide their rations into smaller portions, turning the bread in their mouths before swallowing, in intervals of minuscule nourishment that helped to convince the deep parts of their minds that they were not in mortal danger.

By December bread rations were made with only a little more than 50 percent flour, the rest made up with bulking additives of negligible value. A group of Leningrad scientists attempted to use a hydrolysis process to make cellulose drawn from wood and cotton edible. The processed material added to the bulk and weight of the bread rations, but while it filled people's stomachs, it had no nutritional value and caused abdominal pain. The typical regrets of peacetime and plenty— the loves lost, the opportunities not taken, the words left unsaid—had vanished, replaced by the sorrows of the formerly wasteful. The memory of every plate of leftovers, discarded crust, and overlooked potato peeling tormented the starving.

"I think that real life is hunger, and the rest a mirage," the Russian linguist Dmitri Likhachev wrote of his experience of the siege. "In the time of famine people revealed themselves, stripped themselves, freed themselves of all trumpery." Fixated on his former negligence, Likhachev felt "as much remorse and despair as if I'd been the murderer of my own children." As he lay in bed, the academic thought of one thing "until my head hurt." In his mind's eye he saw tins of canned fish on the shop shelves. "Why hadn't I bought it?" he wrote. "Why had I bought only eleven jars of cod-liver oil and not gone to the chemist's a fifth time to get another three? Why hadn't I bought a few vitamin C and glucose tablets?"

The city's meager rations were insufficient to sustain a human body. As each household's cupboard stocks ran out, the population turned to substitutes that few would previously have considered. Enterprising cooks discovered that the husks of linseed, cotton, hemp, or sunflower seeds, pressed into blocks and typically fed to cattle, could be grated, fried, and turned into a sort of pancake. Some ate joiner's glue, made from the bones and hooves of slaughtered animals, just about edible when boiled with bay leaves and mixed with vinegar or mustard. Neighbors broke into the empty apartments of evacuees and rifled through their drawers and cupboards to salvage every mold-speckled leftover. Anything that contained calories was considered: cough medicine, the powder used for toothpaste, cold cream, machine oil. None of these substitutes, some of which caused gut-twisting side effects, was enough.

"I am becoming an animal," wrote one teenager in her diary. "There is no worse feeling than when all your thoughts are on food." As temperatures dropped and bodies began to shiver and develop brown adipose tissue used to heat the human body, the calories needed to maintain a person's strength increased. Rations, already ruinously diminished, now had to go further to provide the basic energy necessary for survival.

III

STARVATION IS A FIRE THAT consumes from within. When the intake of calories falls drastically short of the energy needs of the body's fundamental systems, the body begins to burn itself as fuel. Carbohydrates, fats, and, eventually, the protein parts of body tissue are thrown into the furnace. Metabolism slows so the body can no longer regulate its temperature. Soon it loses the ability to supply vital organs and tissues with the necessary nutrients. The heart and lungs shrink, reducing a person's capacity for exertion. The mind only permits the body to engage in activities that serve its physical survival. One survivor recorded that she found it "psychologically easier to drag a bucket of water up the stairs than to reach out for a pencil sharpener."

The brain diverts resources to nourish itself for as long as it is able, but eventually it, too, begins to shrivel, a reduction that takes with it clarity of thought and sometimes personality. "[Starvation] gave a person a protective indifference, under cover of which he could die easily," wrote the literary scholar Lydia Ginzburg, who managed, despite everything, to record her experience in its midst. Muscles reduce, contributing to an overwhelming feeling of frailty and exhaustion. The mind flits. Irritation. Hallucination. Convulsion. The heart begins to beat erratically, then not at all.

War left its mark not only as craters beside the road, but as intimate bruises, blotches, and hollows on the skin. Malnourishment robbed the city's citizens of bodily form and sex, turning men and women into equally androgynous specters, their limbs protruding as bones, every gaunt face a stranger in the mirror. "[Whether] sitting, standing, or lying, I am reminded all the time of my extreme emaciation," wrote one academic at the Hermitage, who managed to maintain a sense of humor, even as his body hollowed out. "Especially striking is the disappearance of my buttocks, the one really distinguished aspect of my person, of which I was very proud."

Physical deterioration presaged mental collapse. The mind was the line to be held to avoid complete breakdown. Ginzburg recorded the dispiriting feeling of waking up each morning to a new day, which carried with it the sensation of "being linked into a series of constantly renewed miseries."

There were as many responses to starvation as there were types of starving people. Mothers divided their daily ration into smaller portions, passing it to their children throughout the day to create the illusion of plenty. In one orphanage the children would roll pieces of bread, stowing the tiny balls in matchboxes. "The children would save the bread as if it were the most exquisite food and would savor every morsel, taking hours to eat," recorded one of their caregivers. These wretched techniques maintained a sense of routine, the ritual of a three-meal-a-day structure, which provided a form of psychological sustenance, if not genuine nourishment.

Elderly men and infants soon proved to be the most vulnerable segments of the general population. Starving mothers lost the ability to lactate. "My milk has dried up," one young mother wrote in her diary. Every evening the young mother drank a pot of water. The ritual did nothing. "Lena screams and tears at my breast like a small wild animal. . . . Now we give her all the butter and sugar we get with our ration cards."

Some people, knowing they were dying, still chose to share their bread with one another. Mothers fed their rations to their starving children, even orphans they had taken in. The more one shared, the more one sacrificed. Others lived more selfishly, stealing food, and hunting stray animals. With her mother bedridden, one young daughter set out to collect the day's rations. She took the bread, and as she turned to walk home, a boy snatched the food from her and stuffed it into his mouth. Witnesses grabbed the boy and struck at him while trying to pull the food from his mouth. Seeing his wide eyes and the blood run from his mouth, the young woman implored the crowd to stop their attacks, shouting, "Don't touch him! Don't touch him!"

The primary urge to survive tested morals. In November a teenage boy recorded in his diary how the mother of a friend "called me a complete idiot for not stealing food from our neighbors." She, by contrast, "wouldn't think twice." Still, many of those who survived began to believe that their survival betrayed a failure to sufficiently share and sacrifice.

In the most extreme cases, desperate people resorted to murder. In one hostel a sixteen-year-old machinist was murdered by a workmate after boasting of having exchanged several days' worth of coupons for food. An eighteen-year-old who had lost his job and, with it, his ration card killed his two younger brothers with an axe. He was arrested while trying to kill his mother.

Starved of food and with a gathering surplus of bodies around them, some were tempted to cannibalism. As early as November, meat patties made from ground-up flesh went on sale in Haymarket. Some believed the meat was human, cut from cadavers; traders protested it came, in fact, from horses, cats, or dogs. Others ate people knowingly. Walking through the silent streets one night, a man happened upon the heads of a man, woman, and child, her blond hair still plaited in braids. Their bodies, he guessed, had been carried off for butchering. In the street "one often saw corpses with the buttocks carved out," noted another eyewitness.

When students living in the dormitory of a school were not provided with ration cards for December, they lived for three weeks off the meat of captured animals. Then, on Christmas Eve, a few of the students began to eat the corpse of a classmate who had died earlier from malnutrition. Three days later another student perished; his body was treated the same way. Eleven of the students were later arrested and charged with cannibalism, to which they all pleaded guilty.

The authorities arrested a twenty-four-year-old nurse for, they claimed, scavenging amputated limbs from an operating room. In midwinter Elena Kochina took her daughter to the clinic. She laid her on the table and the pair waited for the doctor to arrive. Presently a nurse approached and whispered, "Don't leave the child unattended. We've seen cases of kidnap."

The city police and NKVD created special divisions of police and psychiatrists to investigate these alleged crimes; at least thirteen hundred Leningraders were arrested for cannibalism during the siege. "We are living the primitive life of savages on an uninhabited island," wrote one diarist.

The scarcity of essentials recalibrated individual priorities and standards. It also rearranged the economy. At the city's central markets people would exchange a gold watch for a fistful of turnips, or a Persian rug for a chocolate bar, an engagement ring for a pound of bread. Those who occupied well-fed positions of importance took advantage of the new economy. A young mother of two small children pinned an advert to a post near the bread shop offering her gramophone in exchange for bread. A soldier turned up at her apartment the following day to make the trade. When the bread was gone, she advertised her precious sewing machine.

"Before long a woman came and offered me slightly more than half a loaf," Okhapkina recorded. "I was reluctant to part with the machine, but I let it go." The woman who took the sewing machine appeared relatively healthy and Okhapkina asked her where she worked. "She told me to mind my own business."

Wants melted and left only elemental needs. Soon there were few valuables of sufficient worth left to exchange for food. A ration card became a person's most precious item, on which one's survival was entirely contingent. In the event of loss or theft, a person would almost certainly die before securing a replacement. To dissuade thieves, military patrols would stop and search anyone they considered to be acting suspiciously. If a stolen card or unaccounted-for food supplies were found on a person, the soldiers, emboldened by the support of the NKVD, would shoot the person in the street. Collective punishment became routine, too. "If they discovered that bread had been stolen," one eyewitness later recalled, "they would round up five people and shoot them for it, whether they had been involved or not."

Scarcity forced the city to rediscover nearly forgotten peasant knowledge and adopt practices long abandoned by the educated and well to do. Leningraders became connoisseurs of wooden fuels, prizing birch above aspen for its warmth-giving qualities; they learned how to roll a cigarette made of dried maple leaves, and how to light the roll-up using sunlight focused through a lens, or from the spark caused by flint chinked against metal. They devised homemade lamps by filling a bottle, tin cup, saucepan lid, or even a spent shell casing with burnable liquid such as camphor, machine oil, or eau de cologne. A wick would be dangled into the liquid; when lit, the device, often known as a *koptilka*, or "smokehouse," would provide a little light and a lot of smoke.

As well as adjusting the value of items, the siege shifted the status of certain professions. Zoologists were well positioned to survive the siege because they knew how to catch rats and pigeons. Bookish academics were more likely to perish. The siege wiped out urbanite sentimentality toward animals. At the Physiological Institute famished researchers ate Pavlov's renowned dogs. Police officers butchered their service animals. An enterprising father brought home the maggot-riddled knee of a reindeer killed in an air raid at the zoo. Pigeons disappeared from St. Isaac's Square. Families, unable to survive on the crumbs salvaged from the grooves of their dining tables, ate beloved pets. Households would swap animals to ease the torment involved or barter resources for cats. Families were naturally squeamish (one employee of the Hermitage knew of a young woman who choked to death when she returned home to find her mother had killed her cat) and would even debate the ethics of eating domestic animals. By winter, few could afford the luxury of moral debate.

"The smallest Leningrad children," one journalist recorded, "grew up not knowing what cats and dogs were."

IV

FOR NIKOLAI IVANOV AND HIS colleagues the loss of Leningrad's domestic animals had secondary effects. Unlike the potatoes, the seed specimens stowed in their metallic containers and arranged in orderly rows in the buildings' secure rooms could withstand the cold. Soon, however, a different sort of threat arrived. Each day Ivanov and Rudolf Kordon would unlock the doors, open the cabinets where the tins of seeds were stacked, and check the condition of the boxes. During one routine inspection, Ivanov was examining the containers when a rat jumped from a shelf onto the floor in front of him. Cornered, it threw itself at Ivanov and bit his leg above the knee. The scientist hurled the rodent off himself, afraid he might catch hepatitis. If vermin could enter the locked rooms containing the priceless collection, the seed bank had a major problem.

The mice and rats came in their thousands. The closure of grocery stores and canteens had made the animals desperate, while the absence of cats and dogs had made them bold. Ivanov and his colleagues built rudimentary traps—large cages baited with scraps—and filled any holes in the walls and skirting with shards of broken glass and dustings of arsenic powder. Each day the scientists would check the cages, which would be filled with writhing, starving rodents, which they removed from the building to bludgeon, unable to eat the meat lest the animals had touched the poison.

The rats soon learned to push the boxes off the shelves onto the floor. The lids and sides of the containers had two round holes, and the insides were covered with thin gauze to allow the seeds to breathe. The rodents chewed at the meshing, widening the openings until they could reach the specimens inside.

Rats even managed to enter the cellar, attracted by the smell of the potatoes. One morning Lekhnovich arrived to check the stocks to find two sacks containing the English variety known as belladonna had suffered "a frenzied attack." It was only after Lekhnovich had bricked up the nest hole that "the battle began to ease."

Upstairs the staff regrouped. Kordon, the Institute's apple expert and chief keeper of the collection, suggested removing the seed containers from the shelves and tying them into bundles, so the meshes were pressed against one another to prevent access. The tied bundles could be hidden under metal roofing sheets, through which even the most determined animal would be unable to chew. Belyaeva, the Institute's building manager, agreed to the plan and took some assistants to collect the metal sheets from the roof. After handling the frozen sheets, the botanists felt as though their fingers had been burned.

Preparing the seeds for long-term, vermin-proof storage proved laborious. Each day a group of ten to twelve staff members, mostly women, took down between three and four thousand boxes from the shelves and tied them together into tightly bound bundles of around ten. At the end of the day, by the light of their kerosine lamps, they would cover these bundles with the tin sheets from the roof and, when they ran out of sheets, pieces of plywood. In cold rooms, with swollen fingers, the work was painstaking. "After the war is over our country will need these seeds more than ever," Ivanov would often say to inspire his colleagues. It was a plea for all to look beyond their own discomfort and fatigue and find focus and hope in a greater purpose.

The weakened team prioritized the wheat collection, stowed across more than twenty thousand boxes, which would be critical for making flour in the future. Then they worked their way through the rye, oat, and barley collections, then maize, millet, sorghum, buckwheat, peas, and fifteen hundred other boxes containing assorted varieties of legumes. Finally, the staff tied up the tins of vegetable seeds and of industrial and forage crops. Seeds gathered from unsorted drawers were put into some two and a half thousand metal boxes. "The entire job was carried out in semidarkness in cold rooms with broken windows," Ivanov recorded. On damp, frosty days, the columns of St. Isaac's Cathedral would gleam from hoarfrost, which also coated the metal boxes in the seed bank, lighting up the rooms.

When the task was complete, the staff had tied together a hundred thousand boxes, which had been spread across forty of the Institute's rooms. Now, fearing that hunger might drive some of Leningrad's citizens to follow the rodents in storming the seed bank for food, Klavdiya Panteleyeva arranged for the wearied staff to gather the bundles into rooms on the second floor, away from street level and easy reach of trespassers. Once finished, the staff took a fresh inventory and drew up a layout of the new storage arrangements, to use during their once-weekly inspections.

Finally, Panteleyeva and Kordon locked the rooms and sealed the doors. Unable to reach the seeds, the rats began to chew loose documents to shreds and gnaw at the wooden legs of the scientists' desks. Slowly their numbers began to decline until, finally, there was no evidence of any vermin inside the building.

V

THE EVACUATION OF MICE AND rats only marginally reduced the challenges faced by the beleaguered staff who remained at the Institute. As trams began to run more infrequently, then not at all, some of the department heads moved into the building to avoid having to expend valuable calories traipsing through snow from their homes. Every man and woman was still required to put in fire-watch shifts, extinguish incendiaries that landed on the Institute roof, and continue their research duties. Likewise, each day everyone had to help clear the snow and debris from the road outside, which remained a crucial thoroughfare for military vehicles. Even though his legs had begun to swell from the effects of malnutrition, Ivanov was also responsible for fetching water from the nearby Neva River.

Routine provided the gift of a daily focus and purpose. The absence of purpose could be as deadly as cold and hunger. "For many people a regimen, a work routine, was an unattainable dream," wrote one resident. Without a meaningful focus, beyond the most basic concerns for survival, "the effort to set their life in order just wouldn't come." Lekhnovich understood the mortal value of his task to keep alive the potato collection stowed in the seed bank's basement. "During the blockade, people died not only from shells and hunger but also because of the aimlessness of their existence," he later reflected. "In the most direct way, our work saved us. It invested us in living."

The seed bank staff also drew strength and motivation from their absent mentor, Vavilov, whose whereabouts were still unknown, yet whose presence was keenly felt in every office room and corridor. "It is better to display excessive concern now, than to destroy all that has been created by nature for thousands and millions of years," Vavilov had said before his arrest. No monetary price could be placed on the Plant Institute's treasure; its value was not abstract but elemental, and as such, those entrusted with its safekeeping were not motivated by profit but by salvation. The abject need all around the seed bank's staff pressed home the importance of their task: to guarantee food security if not for themselves, then for those who followed them.

While his colleagues worked to protect the seeds from looters, Vadim Lekhnovich considered his options to ward off the murderous frosts that threatened the potatoes. Finding fuel to heat one's room was proving almost as difficult as securing food, however. In September the last two-and-a-half-liter ration of kerosene had been distributed to Leningraders. In late November residential buildings were banned from using electricity between 10:00 a.m. and 5:00 p.m. Even during permitted hours, power was supplied erratically or, more often, not at all. In St. Isaac's Cathedral, museum staff filled the altar lamps with seal fat obtained from the zoo.

Residents turned to *burzhuiki*—small stoves that took their name from *bourgeois*, referring to their potbelly appearance and greedy fuel consumption—with trailing flues that could be angled out of a cracked window. To plug drafts, citizens would wedge pillows or blankets in the gaps around the chimney. "*Burzhuiki* have become the hottest commodity," wrote A. B. Suldin, on November 15, noting that many used the stove to toast their ration of blockade bread, the easiest way to rapidly dry out a foodstuff that now came as "wet as clay."

When their firewood ran out, some Leningraders began to burn books, furniture, parquet flooring, and even dried-out feces. Alexandra Dyen and her son Vladimir removed the wooden shelves from the kitchen cupboards for burning. When the shelves were gone, they burned the kitchen table. The wardrobe kept them warm for three weeks. Finally, they turned to the family library.

"I burned the German classics, and after that it was Shakespeare," Vladimir later remembered. "I also burned Pushkin . . . I think the Marks edition in blue and gold. And into the fire went that well-known multivolume edition of Tolstoy's works." Families hunched around the stoves, dandling invisible babies on their knees, trying to heighten their sense of warmth.

Outside the city, German officers issued their troops guidelines on how to keep warm: a lining of newspaper slipped between layers of clothing to protect the lower abdomen; a handkerchief or balaclava to line the helmet, arm warmers fashioned from old socks. A lucky few wore fur coats looted from homes abandoned by residents en route to Leningrad. Most troops tied their heads with women's scarves, wore children's bonnets, and filled their boots with straw.

Twice a week Lekhnovich heated the storeroom in the main Institute building. "Only occasionally did the temperature in the storeroom fall below zero," he recorded. Maintaining a steady heat in the cold, dark cellar where he had planted the bulk of the rescued specimens was another matter, however. He first plugged drafts with cotton wool, sacking, and rags. Then he moved a stove into the cellar, hoping to raise the temperature by a crucial, crop-saving couple of degrees.

Each day Lekhnovich lit the stove, which threw out blooms of choking black smoke. The stove burned through fuel and cooled almost as quickly as it flared up. The regular allocation of firewood was not enough to warm the cellar. If Lekhnovich was to save the potatoes, he needed a supplementary source. After he returned home under pale and sullen skies, he covered himself and his sick wife with a weighty pile of blankets and coats and mulled over how he might increase his fuel ration.

Once a week the Plant Institute's warden, Maria Belyaeva, supplied Lekhnovich with a bundle of firewood, which he divvied into daily quotas. He would sit and listen to the grudging crackle of the fire. Then, when the day's ration ran out, he would burn whatever scraps he could find: old boxes, cardboard, paper, and burnable debris scavenged from bomb-wrecked buildings, anything that might raise the temperature in the dark cellar by a couple of degrees. On one scavenging trip he approached a group of soldiers guarding the hospital, who were warming themselves by a bonfire in the yard. They offered him a drawer from some old table or sideboard—a piece of refuse turned to treasure in the oppressive cold.

Each of the remaining seed-bank staff did their bit to bolster the shrinking supplies of wood, too. Ivanov's daily commute now took the botanist an hour on his swollen legs. His progress was hindered not only by fatigue, but also by the weight of a hammer, with which he knocked pieces of wood from the debris of flattened houses, to be used as fuel at the Institute. On a good day—if plundering the wreckage of his neighbors' homes and places of work could ever be considered a good day—Ivanov would arrive at work with twenty-eight pounds of firewood.

Soon, however, Lekhnovich found he no longer had the strength to lift the wood. He had seen other residents using *sanki*—child's sleds—to transport fuel. With no sled to hand, he loaded the wood onto a sheet of plywood and dragged it across the square, sitting on the bundle to rest at regular intervals.

One day, after the seed bank ran out of fuel, Belyaeva handed her colleague a requisition from the October Region Executive Committee for half a cubic yard of firewood. Lekhnovich headed across the city with his improvised sled. He approached the elderly man guarding the truck on which bundles of damp pinewood awaited collection. As Lekhnovich drew close, an artillery shell hit the corner of a nearby building. Both men ducked as pieces of shrapnel howled past. The guard cursed and moved to take cover.

Lekhnovich stopped him. "Let me take the firewood. Here's the requisition note."

"Can't you see what's going on?"

"I need that firewood now," Lekhnovich insisted.

The guard stared at the young botanist with the straggly white beard. After a moment he waved Lekhnovich toward the truck, then retreated to cover. In the empty street, under skies muffled with the threat of artillery fire, Lekhnovich began to lay the logs out on the plywood board. When he estimated he had the correct amount, he began to drag his makeshift sled back toward St Isaac's Square. Each time he heard the scream of an incoming projectile, he knelt to the ground and held a piece of firewood above his head as a form of pitiful cover.

Numb in the snow, the botanist smelled the sweet odor of tar that had soaked into the wood. Then, delirious with defiance and a sense of wild, mortal purpose, he began to laugh.

During the winter of 1941, stacks of corpses several yards tall and thirty yards long
rose in the cemeteries and hospital courtyards. Corpse collectors drove hundreds
of bodies to a huge wasteland on the outskirts of the city. Local defense units used
explosives to churn up ground, then lowered bodies into the mass graves, which
eventually and collectively became known as the Piskaryovskoye Cemetery.

IX

A SILENCE OF ANGELS

I

On December 11, 1941, the day that Adolf Hitler declared war on the United States, Andrei Zhdanov, Stalin's rotund, chain-smoking lackey, joined a clutch of leaders at the Kremlin in Moscow to discuss the plan to liberate Leningrad. The son of a schoolteacher, Zhdanov had risen through the Party ranks to become responsible for almost every facet of municipal life in the besieged city. A reluctant delegator, his responsibilities had weighed ever more heavily since the invasion; Zhdanov's staff joked that not a volt of electricity was allocated without his consent. But his burden was psychological, not physical—aside from the dusting of dandruff on the shoulders of his uniform and regular fits of asthmatic wheezing. While Leningrad's citizens butchered their pets for food and burned their bedtime reading for warmth, Zhdanov lounged in his study eating wedges of buttered white bread topped with black caviar, and drinking strong sweet tea, while his staff forlornly watched on.

The atmosphere at the Kremlin meeting was optimistic. Russia's losses were tremendous and devastating: close to three million Red Army soldiers, including its most competent and best-trained troops, had been killed since the invasion. This was an unseeable number, each life a tragedy with its own texture and reverberations of impact (the missing lover, the absent father, the life unlived) reduced to a lifeless statistic on a military accountant's spreadsheet. A million and a half more were wounded. And yet hunger and the ceaseless miles and depthless cold of the Russian winter had drained the German army's resolve, too. Those who had not frozen to death while curled in foxholes had lost fingers to frostbite or gangrene. The boots of some German soldiers who were unable to walk had to be cut off to enable medics to discern whether the frozen limbs inside could be massaged back to life or required amputation. German morale was lost in the brutal expanse of the Russian topography.

The Nazi assault on Moscow, code-named Operation Typhoon, which had begun at the end of September, had all but petered out. First it had been slowed by the icy sludge on the ground, then repelled by Marshal Zhukov's well-orchestrated resistance (incentivized by Stalin's persuasive threat to kill him should the city fall). Soviet troops had begun to recapture some key strategic points. "The triumphalism with which we began our advance on Moscow has completely evaporated," wrote SS *Hauptsturmführer* Herbert Lange of the Fifty-Sixth Infantry Division a week later. "We are losing our self-confidence and self-belief."

Across the front, German divisions had begun to retreat, withdrawals too sizable and chaotic for any officer to dissuade with the wave of a pistol. Hitler was also at war with his generals, whose early successes had first slowed, then stopped; most would be replaced within weeks.

At the Kremlin, Zhdanov seized the moment to petition for his city, which had been all but institutionally abandoned since the invasion began. For months Stalin's focus had been the defense of Moscow. For him Leningrad's primary role was to keep its war plants churning out the munitions required to hold the line outside the capital, irrespective of the siege situation. Zhdanov carefully explained the heavy toll of air and artillery bombardment on Leningrad, and the cost of hunger and cold on its inhabitants. Together these forces had contributed to the deaths of more than eleven thousand citizens in November, and at least that many again during the first two weeks of December. Flour stocks would run empty before the end of the month. The situation, by every measure, was desperate.

The response, Stalin's chief of staff, Marshal Boris M. Shaposh-nikov, suggested, should be a three-pronged attack: first to drive the Germans from the territory east of the Volkhov River, then to continue onward to break the grip of Army Group North's siege forces. The mood in the room was hopeful. Buoyed by the prospect of a change in fortunes, and perhaps aware that, following thousands of deaths, there were fewer mouths to feed, on the return flight to Leningrad, Zhdanov decided to boost the winter ration: a seasonal gift of an extra seven ounces of daily bread to every surviving man, woman, and child, the equivalent to four slices.

II

At one o'clock in the morning of December 25, 1941, Leningrad's Party workers received news of the increased ration. That morning there was almost no electricity in the city, which was by now served by a single power plant, the Red October. The bread stores opened in five hours' time. With the radio system down, starving Party representatives were unable to communicate the news remotely, so set out on foot into the dark and frozen city. The message risked being lost before it could be enacted. "On my way to Vasilyevsky Island," recalled one the messengers, "I had to stop and rest five times."

The increase offered an unexpected benison for a population that was in no state to celebrate the holiday. For one eight-year-old boy, the box of Christmas tree decorations was no longer a trove of delightful trinkets but a potential source of food. He and his sister rifled through the box searching for last year's walnuts. "Their insides were dry and shriveled, but we ate them," he recorded. "I can't say that it cheered us up, it was just a way to pass the time."

News of the ration increase spread slowly from person to person, but joyfully. On her way to the bakery, one young nurse met a man who was crying and laughing while banging his head. She decided he must be drunk or driven mad with hunger. When she reached the bakery and learned about the ration increase, she realized he had been delirious with happiness.

Word spread, and with it hope. Before the New Year, the story went, Mga would be retaken, reopening supply lines. Trains, their carriages scrawled with the words *Prodovolstviye dlya Leningrad*—"Food for Leningrad"—laden with unthinkable feasts, would soon follow. For many of those in the advanced stages of starvation, the good news came too late. At least five thousand Leningraders died on December 25. On his commute Lekhnovich passed the bodies of those who had fallen where they walked, who had apparently died unmourned or unnoticed, and on whose bodies the snow had fallen, and with it a silence of angels.

III

BY DECEMBER ALEKSANDR GAVRILOVICH SHCHUKIN'S health was failing. The botanist's skin had dried and blackened, his nose sharpened. When his hands became swollen and imprecise, he no longer shaved, fearing he might cut himself and never stop bleeding. Twice Shchukin fell while climbing the steps to the front doors of the Plant Institute. A doctor wrote him a note excusing him from work. Shchukin refused to take it. There was too much to be done.

Fifty-eight, shy, and polite, Shchukin had studied, lived, and worked in Leningrad his entire life, becoming an associate researcher of the collection of industrial and forage crops at the seed bank, and an expert in groundnuts. In 1921, during the height of the Civil War famine, he published a pamphlet on rabbit farming, arguing that this was a straightforward way to guarantee a supply of cheap meat. He knew how to improvise in a catastrophe. Since the invasion he had steadied himself with ritual and routine. He would arrive at the seed bank promptly each day, five minutes before he was due to start work. He would hang his overcoat and galoshes in the cloakroom, then disappear into his study until lunchtime. Before his abduction, Vavilov wrote in Shchukin's staff file, "Works slowly, but extremely accurately."

Still, in the depths of winter it had become almost impossible for Shchukin and the other botanists to maintain their work schedules. Herzen Street was a strategic route along which Red Army troops and munitions required constant access. Responsibility for maintaining the road outside the building fell to the seed bank's staff. Each day those with the strength to do so cleared snow and garbage, collected melt-water in buckets and poured it down nearby wells, and broke up and removed ice. Once a week the Institute's roof had to be cleared of snow to ensure it did not compress into slabs that might slide off and injure pedestrians or vehicles in the street below. The staff had to carry piles of snow by sled or cart to the Moika River. These demanding chores became increasingly difficult for the emaciated staff, who had, by mid-December, almost reached the end of their physical capacity for labor. When this snow-clearing work was done, few had the energy to return to their scientific endeavors.

On December 25, just as news of the increased ration spread, Ivanov went to check on Shchukin, whom he had not seen for several hours. Ivanov opened the door and gasped. Shchukin was seated motionless in his desk chair. Ivanov ran to his friend and shook his shoulders. But Shchukin's body was already stiffened and cold, one hand locked in place on his chest. As Ivanov attempted to loosen Shchukin's arms to angle them by his sides, a packet of almonds fell onto the desk. Shchukin had died while clutching specimens that could have extended or even spared his life.

As Ivanov wrote, Shchukin was "the personification of a hard worker: efficient, honest, polite, and demanding of himself and others." As she typed up the death notice, Nadya Katkova, a young laboratory assistant in the department of cereals, wept for her colleague. Her grim work, however, was just beginning.

IV

DEATH WAS NO LONGER A mere unwelcome intruder who periodically upset the rhythms of life in Leningrad. Death had become an obstinate squatter, from whom there was no psychological respite. "Death is now a common occurrence; we are used to it," wrote one teacher. At night, Lydia Ginzburg recorded, people battled for life like "perishing polar explorers." Those who survived to see the dawn found "the cold that was to torment them through the day lay all around." (The well-read survivors, Ginzburg noted, hunched in their rooms recalling the lowest circle of Dante's hell, frozen over by the flapping of Satan's wings.)

"Corpses in the gateways, corpses on sledges, lanky and thin, more like mummies than normal human corpses," Ginzburg wrote, of death's work now spilled into the streets. In some parts of the city, cadavers became as commonplace as litter—at least one bundle of expired humanity per hundred-yard stretch of road. According to reports of the German Security Service, these individuals had collapsed through exhaustion, then, with no one willing or able to help them back to their feet, had frozen to death in the snow.

The writer Elena Skrjabina described Leningrad as being "literally flooded" with corpses. "Death reigns in the city," she wrote. "Relatives or friends take them to be buried, tied on by twos and threes to small sleds. Sometimes you come across larger sleighs on which the corpses are piled high like firewood and covered over by a canvas. Bare, blue legs protrude from beneath the canvas. You can be certain this is not firewood."

Propaganda failed to convince anyone. The streets had become both hospital ward and morgue; in the first ten days of 1942 more than two thousand people were picked up off the streets, alive but unable to move, and sent to hospitals; another thousand dead bodies were removed. Skrjabina noted that bereavement had become so commonplace it no longer aroused in her pity or emotion.

One time Skrjabina watched a man shuffle in the snow, then sit down heavily on a fire hydrant. She watched the eyes in his "blue, cadaverous face" roll back. Then he slipped to the ground. By the time she reached him, he was gone. Another man in his forties queued for two hours for food, before handing over his ration card in exchange for two portions of soup and porridge. He ate the soup in the canteen, but left the porridge untouched. The bowl soon attracted the interest of those around him. A waitress approached and tried to rouse the man, only to find he had died at the table, midmeal. As they awaited a policeman to take care of the situation, the crowd did not disperse, but loitered, each person hoping to be the recipient of the uneaten bowl.

In December 1941 Leningrad's officials recorded 52,612 deaths, almost exactly as many people as had died during the entire previous year. In the first ten days of January, more than half that number again died. Years earlier, a Russian diarist had written, "To die in Russia in these times is easy, but to be buried is very difficult." Never was this truer than in the winter of 1941, where the physical cost of breaking the frozen burial ground was often too great. Stacks of corpses several yards tall and thirty yards long rose in the cemeteries and hospital courtyards. Corpse collectors drove hundreds of bodies to a huge wasteland on the outskirts of the city. Local defense units used explosives to churn up ground, then lowered bodies into the mass graves, which eventually and collectively became known as the Piskaryovskoye Cemetery.

With each death the burden on those civil servants responsible for feeding Leningrad's citizens was eased at a dividend of nine ounces per day, lending credence to Stalin's gallows quip that "death solves all problems; no man, no problem." But the optimism felt during the meeting of Soviet leaders at the Kremlin in December had failed to blossom into tangible military gains. Red Army troops, weakened, sickened, and low on morale, had neither the physical strength nor the munitions necessary to dislodge the Germans from their positions around the city.

With death evident all around them, those inside the siege ring felt despondent. Stalin's focus was on fortifying Moscow, not redeploying troops to attempt to break the siege. Throughout the winter months Leningrad's factories transported vital weapons and ammunition to the Russian capital via plane across Lake Ladoga, compounding the sense that, even in the city's greatest moment of need, Leningrad was expected to give, not receive. The people felt isolated and abandoned; as the state controlled the media, citizens elsewhere in Russia remained broadly ignorant of the specifics of their suffering.

Despondency often turned to suicidal thoughts. An artist known as ONOPKOV saved up half a cup of peas, a piece of ham, and enough tobacco to fill a few pipes. On January 15, 1942, he made dinner, smoked his pipe, then injected himself with morphine. "I cannot allow my wife and mother to watch as I slowly die," he wrote. The head of museums in the city's Department of Cultural Affairs proposed he and his staff make a gruesome pact. "We shouldn't wait for a slow death from starvation, but instead gather enough courage and kill ourselves," he suggested. After all, "in the spring the Germans will take Leningrad."

By the end of the year many of Leningrad's citizens had lost the use of their hands, their fingers locked in rictus uselessness, unable to grasp, only their thumbs able to move against the malnourishment and cold. Having realized that death from starvation was now a likelier fate than death by shelling, few sought shelter when the air-raid sirens blared, preferring to continue their labored mission to the bread store to collect their rations. To expend one's strength in search of shelter had become an expensive indulgence. As they retired to bed, some Leningraders would disconnect their radio so they could no longer hear the air-raid warnings. After December 17, the sirens fell completely silent: the German planes had been grounded by the cold, which caused their fuel to freeze.

While a few dozen Russian aircraft succeeded in delivering food supplies each day, these deliveries fell far short of the requirements to feed even a fraction of the three million people trapped in the city. For those inside Leningrad, careworn and blinded by hunger, which had become the dominant sensation, hope now failed on a collective scale.

"There is no record in history of a city with a population of a million being in a state of siege for a fifth month," wrote one engineer. "There will come a point soon that all of us living in Leningrad will perish from starvation," wrote another factory worker, whose diary was confiscated by the secret police. In the brutal new year of 1942, the mortality rate would soon average the appalling height of 389.8 people per 1,000; two in five residents were dying.

Despite the harrowing abundance of evidence, Soviet news broadcasts would admit only to "hardship" and "shortage," never "starvation."

A decade earlier the Stalin administration had criminalized the use of the word *famine*, to avoid the implication that the government's collectivization policies might have caused the catastrophe. Now, in Leningrad, language was similarly used to deflect reality. Institutional deceit was immortalized in death certificates. Leningraders, according to the authorities and those they made complicit in the lie, did not starve to death but rather perished of *distrofia*—dystrophy—a medical euphemism coined by Soviet doctors to reposition mass starvation not as a municipal deficiency, but as a deadly disease.

The word was precisely chosen to divert responsibility for the death of Leningrad's population away from the state, which was unable to provide the minimal nourishment required to prevent the fatal wasting away of muscles and mind. "Dystrophy" manifested grotesquely on the human body: stomachs hollowed; women ceased to menstruate; facial features melted; skin became taut and blackened; speech slurred as vocal cords atrophied; lips retreated until they no longer covered one's teeth, causing saliva to drip ceaselessly. In their diaries, citizens recorded the strange sense of alienation from their bodies, rendered unfamiliar in contour and sensation by a chronic lack of food.

Others working at more favored government-affiliated institutions in the city fared better. Communist Party personnel constituted an elite group who enjoyed unthinkable privileges. According to one account, Cafeteria No. 12, situated in the Party headquarters at Smolny, east of the Plant Institute, along the Neva riverbank, provided not only bread but also sugar, cutlets, buckwheat, and small pies well into the winter weeks. One artist recorded in her diary how she visited the public baths and was "completely astounded by the large number of well-fed Rubenesque young women with radiant bodies and glowing physiognomies." These women were all, she noted, workers in bakeries, cooperatives, soup kitchens, and children's centers.

Evidence of the desperation outside the Plant Institute's hushed and frosty corridors was clear to the botanists, however. One morning Lekhnovich arrived to check the potato collection and noticed the cellar was unusually cold, even though the doors, locks, and seals appeared intact. He entered the room and checked the thermometers, which registered a temperature of minus one at ground level, and zero degree at a height of three feet. Holding his night-light aloft to see in the dark, he checked the stonework.

"The source turned out to be a narrow chink at ceiling level," he recalled. A burglar had fashioned an opening into the cellar from the building next door and, using some kind of tool, had hooked several sacks of potato hybrids from the top shelf. Dismayed, Lekhnovich summoned his colleagues, and together they filled the hole with old sacking and lit the potbelly stove to warm the potatoes.

"No point in fixing an ordinary lock: they will just crowbar it off," the botanist reasoned. Instead Lekhnovich found a five-foot-long piece of piping from the scrap heap and used this to jam the cellar door from the inside. "Then, with some difficulty, I squeezed through the chink, which I then closed up." When he arrived at work each day, he checked that the door was still sealed. Whenever fresh snow fell, he sometimes saw the footprints left during an attempted burglary. But no one successfully prized open the door. The specimens remained safe inside the cellar.

V

THE WEEK THAT SHCHUKIN DIED at his desk, a German shell killed Yevgeni Wulf, head of the Institute's herbarium. Shortly afterward Alexander Moliboga, a senior researcher, died in a fire caused by a bomb that landed on his apartment; he had been too weak to flee the resulting blaze. With news of each new death, the gloom enmeshed itself more deeply into the fabric of the Institute building. In early 1942 temperatures continued to drop, reaching record lows of between minus -36°C and -40°C, at which point hypothermia sets in after no more than seven minutes. The cold caused a deadly run of winter weeks. The Plant Institute became a mausoleum.

On January 5, eleven days after his colleagues found Shchukin's body, Dr. Nikolai Petrovich Leontievsky, a senior researcher at the Plant Institute's agrometeorology department, died of starvation. Like

Shchukin, he had volunteered as part of the firefighting group for the seed bank, helping to extinguish incendiaries that landed on the roof.

Four days later, Dmitri Sergeyevich Ivanov—namesake to Nikolai Ivanov—a senior researcher and the head of the rice section, died of starvation. Having served as a divisional engineer in the Red Army during the Civil War, for which he had been awarded the Order of the Red Banner by the Revolutionary Military Council of the Republic, Dmitri Ivanov was a practical man. He headed up the unit responsible for firefighting and general defense of the building at 44 Herzen Street. Like Shchukin, he died in his office, surrounded by his research papers and several thousand packets of rice samples.

Weeks earlier the botanists had collectively decided they would not consume the seeds in the collection. Now, everyone's commitment to the plan and resolve to uphold it was privately tested to the extreme. In thousands of tins sat packets of nuts and seeds that could be tapped into open palms and consumed by the handful. Almost everything was edible, and the quarter of a million seeds could have been eked out to sustain the botanists for months.

And yet, to consume these specimens would, for the scientists, feel like a betrayal of the past two decades' worth of effort, the thousands of miles that Vavilov had traveled around the world, his gentle convincing and bartering to secure the samples. As employees of the world's first seed bank, the botanists were the only people to have been faced with this ultimate and fundamental dilemma: to save a collection built to eradicate collective famine, or to use the collection to save themselves.

"It wasn't difficult not to eat the collection," Lekhnovich later said. "It was impossible to eat this, your life's work, the work of the lives of your colleagues."

And yet, the costs were tremendous. On January 16, 1942, shortly after his forty-seventh birthday, Georgi Viktorovich Heintz, head of the Institute's library, died of hunger. One of the founders and originators of the Institute's library, Heintz, who had studied in Germany at the Berlin Botanical Garden and Botanical Museum, had developed a subject catalog later adopted by other scientific libraries. Heintz was a polyglot with unrivaled experience. He had worked previously as chief librarian of the Leningrad section of the State Publishing House, Gosizdat, and at the science library and museum of the Nikita Botanical Garden in Yalta. Vavilov headhunted Heintz, believing it was a role for which he was born: cataloging and organizing literature at the greatest seed bank yet assembled. He was buried in the mass communal grave at the Piskaryovskoye Cemetery.

Others followed. That same month the Plant Institute lost Lydia Rodina, keeper of the oat collection, who also died of hunger in the room where the oat collection was packaged for safekeeping. And Dr. Georgi Kovalevsky, the Institute's oldest expert, who worked on the agricultural development of highlands and had authored more than fifty scientific papers on his research; and Nikolai Likhvonen, the Institute's procurement agent, who ran the supply division; and Anisiya Malgina, a close colleague of Heintz's, and head of the Institute's archives.

Around the same time, Professor Samuil Egiz, doctor of biological sciences, head of the tobacco and tea group, and author of more than fifty publications on genetics and tobacco breeding, died of hunger. So, too, did Andrey Baikov, the Institute's driver and mechanic, who had accompanied Vavilov on his expeditions to Transcaucasia and other republics of the USSR in the midthirties, and who had, since the invasion began, assumed responsibility for repairing artillery and fire damage to the seed bank buildings. Each death underscored the stakes and added to the burden of responsibility to those who remained, not only to continue the work of their colleagues, but to follow their example.

For ordinary citizens survival meant enduring feelings of inadequacy, remorse, and, even, a sense of shame. The botanists were shielded from such feelings, somewhat, by their sense of duty and responsibility. Decades later, when asked why he and his colleagues chose not to distribute and eat the seeds to save the starving, Nikolai Ivanov replied, "Because our task was to save the collection. We knew that later, after the war was over, our country would need those seeds more than ever."

There were still some moments of grace and levity amid the mourning. One midday the Institute building manager, Belyaeva, located Klavdiya Panteleyeva.

"You'll never guess what I've found in the storeroom," Belyaeva told her friend, leading her into one of the basement areas. Belyaeva lit a match and used to light the end of a twisted piece of newspaper. She crouched in the corner of the basement, where the wriggling light revealed a pile of kerosine lamps.

"So what?" said Panteleyeva.

Belyaeva picked up one of the lamps and lit the wick with her paper torch. The lamp flared up, bringing its warm light to the entire basement, and revealing three large metal canisters, each one full of kerosene.

"Just look," said Belyaeva. "They're full!"

That evening lamps burned throughout the Institute, as if to mark a winter celebration. The light had a transformative effect on the group, who huddled together in conversation in ways they had been unable to for weeks. Time, one participant recorded, seemed to pass more quickly.

On another occasion the staff members received a package sent from the front by Dmitri Brezhnev, a tomato expert who had run the department of vegetable crops at the Institute until he was called up. Brezhnev occasionally sent letters to his colleagues, but this was different. He had managed to deliver a small crate that Panteleyeva had to drag into the Institute using a sled.

At the front desk, she carefully opened the lid of the crate with a knife, easing away the nails. She pushed her finger inside and pulled off the lid. Panteleyeva cried when she found meal rusks and a few lumps of sugar inside, an unimaginable treasure during the depths of the winter. She divided the provisions equally and brewed some tea for her and her colleagues. Scalding themselves, they drank the tea *vprikusku*—holding a piece of sugar in their mouths, where it sweetened the liquid as they sipped.

"As soon as the war is over, the first thing we'll do is gorge ourselves on bread," said Galya Lebedeva, a lab technician. "I'll cover it in a generous sprinkling of salt and sliced onion. And I will eat."

VI

ON JANUARY 8, 1942, IOSIF Orbeli, director of the Hermitage Museum, received a bleak request from the Union of Architects and the Museum of Ethnography. Both institutions wanted the Hermitage to use any packing materials left over from the evacuation of the museum's works of arts to build coffins for their dead. The request would have been unthinkable just six months earlier. During the summer of 1941, Orbeli had been able to swiftly organize the manufacture of crates and dispatch more than a million artifacts. Now he was unable to manage the construction of a single coffin. The museum's resident carpenter had died. Even if he had lived, nobody in the Hermitage basement could spare the strength. Instead, whenever a person died, museum employees carried the corpse outside, into St. Isaac's Square, to be collected by one of the military trucks that rounded up the bodies every few days.

In bedrooms, kitchens, and living rooms around the city, the bodies of the dead awaited burial. In late December, with the help of her family, the writer Elena Skrjabina had carried an elderly neighbor from the kitchen, where she lay on the table, into a bedroom, and wrapped her in shawls and blankets. She survived for several days, mumbling Estonian prayers. One day Skrjabina fed her sardines; an hour later the woman died. Her body lay in the bed for three weeks before she was finally collected to be buried. Trauma compounded trauma. In the first twelve weeks of 1942, ninety-eight orphanages opened in the city.

Conditions in the museum were difficult. While the seed bank's population was tiny, more than two thousand people had moved into the Hermitage basement. The museum's employees and their families were joined by members of Leningrad's artistic and scientific communities. In the gloaming light and dead air, scientists attempted to conduct experiments and write up their results by candlelight, using ink that was close to frozen. Artists produced new works to make sense of their reduced existence. Despite these efforts to maintain a sense of routine and normality, the cellars were filled with the stench of decay and rot. Stiff corpses lay on bunks where the people had died. It felt, according to one survivor, as though "the souls of the dead were hovering through the vaulted corridors."

A correspondent for *Komsomolskaya Pravda* wrote in his diary, "The city is dying as it has lived for the last half year: clenching its teeth."

VII

AMID THE GRIEF, CLAUSTROPHOBIA, AND horror of this winter, when the forces of nature allied with the invaders, the botanists at the Plant Institute had largely forgotten the freight train carriage that, during the warm and relatively easy summer months, they had filled with a hundred thousand specimens and dispatched to the mountains.

After failing to pass through Mga, the freight car, loaded with some of the most valuable seeds, equipment, and books in the Plant Institute's collection, had been shunted back and forth along the Okty-abrskaya Railway line for six months. As autumn turned to winter, the material had been placed in sidings, moved on by railway workers whenever the enemy's bombs began to close in.

The seeds were well protected, housed in the double-walled boxes that shielded them from the elements, but by the New Year any hope that the material could reach its intended destination had vanished. The surviving staff at the seed bank received official notice from representatives of the Oktyabrskaya Railway that the evacuation had failed; the three hundred containers Vavilov's team had diligently sorted, packaged, and itemized with supporting documentation had to be collected.

With the staff weak in strength and few in number, Nikolai Ivanov replied that it was impossible for his team to meet this request. Andrey Baikov, the Institute's driver, was dead; so, too, were many of the people who might have been able to lift the crates and boxes onto alternate modes of transport. Still, the railway's administrators replied, the crates had to be returned.

Panteleyeva volunteered to visit the offices of the Oktyabrsky District Party Committee. There an official told her that, if the botanists could unload the crates from the train and ferry them to the depot, they would arrange for a tram to carry the seeds to St. Isaac's Square. Panteleyeva explained that nobody at the seed bank was able to unload such heavy boxes. The official took pity and ordered a group of soldiers to assist.

While Panteleyeva supervised, the soldiers unloaded the seed bank's crates from the train and carried them to the tram depot through gales of freezing wind. When they arrived, however, no tram was in sight. Panteleyeva despaired. Then, from around a corner, a freight car rolled into sight. The botanist ran forward onto the tracks, waving her arms at the driver to stop.

"You crazy woman!" he shouted, pulling on the brakes. "Do you want to get run over?"

Panteleyeva explained the situation, but the driver looked bemused. Why, amid famine and war, would anyone want to transport books and seeds? Besides, he explained, he did not have the proper authorization.

"But it's bread that will perish," Panteleyeva explained. "Do you understand? They're live seeds and they're freezing here."

At the mention of bread, the driver's attitude changed. He motioned for the soldiers to load the crates.

When the freight car pulled into St. Isaac's Square, Ivanov and his skeletal colleagues met the soldiers at the entrance to the Institute. The soldiers agreed to carry the crates up the steps and deposit them in the entrance hall, but they would take them no farther. When the last of the crates was off-loaded, Ivanov and his friends looked helplessly at the tons of plant material. The men and women had no strength to lift such heavy containers, and besides, they were sufficiently well packaged to protect their cargo from rodents or thieves. The seed bank staff shuffled off, returning to their offices. The crates would remain unopened until the end of this war, whenever it might be.

In an office room Nadya Katkova compiled a report on staffing. Between June 1, 1941, and New Year's Day 1942, the Plant Institute, and its two experimental stations on the outskirts of the city, had lost ninety-eight staff members. Forty-three had willingly resigned, seventeen had been made redundant. Twenty-one had gone to the front to fight. Seven had left for undisclosed reasons, and ten had died, most to causes related to starvation. Few believed new hope would accompany the new year.

Vavilov (*pictured*) and the botanists at the Institute pledged to never eat the hundreds of thousands of seeds held there. As employees of the world's first seed bank, the botanists were the only people to have been faced with this ultimate and fundamental dilemma: to save a collection built to eradicate collective famine, or to use the collection to save themselves.

X

DO NOT FORGET MY NAME

I

NIKOLAI VAVILOV STARED AT HIS reddening fingers splayed on the asphalt as his knees and hands grew numb. He was not, he had been told, to look up, but still he sensed the damp, wheezing presence of thousands of prisoners all around him. The men, mostly educated Muscovites, were hunched on all fours on the ground. Soviet guards patrolled the scene while their attentive dogs tugged at their leashes and sniffed. Freezing rain dribbled from the leaden night sky onto Moscow. Any man who attempted to sit or shuffle his limbs out of one of the larger puddles was kicked until he returned to the kneeling position.

The first snow had fallen the previous day. By the time Vavilov and his fellow prisoners from the Butyrki Prison arrived, it had melted and left a layer of dirty slush, now hungrily soaked into the prisoners' clothing. It was a miserable scene, as each man trembled in the cold, and awaited his fate.

Since being handed the death sentence in July 1941, Vavilov had languished in prison. After the brutal interrogations of the summer had ended, he had spent his time writing, and arranging a schedule of lectures among his cellmates. He invited each man to deliver a talk on his area of interest or expertise: history, biology, the logistics of the timber industry—whatever it might be. In turn the men held forth in whispers, eager to be heard by their fellow prisoners, but not to alert a passing guard.

Vavilov had been condemned to death but believed that he might have his sentence commuted from execution to imprisonment if he could convince the NKVD that his expertise could prove useful to the war effort. Unbeknownst to Vavilov, the man with the power to overturn the sentence, Lavrentiy Beria—one of Stalin's most senior chiefs and head of the secret police—was the same individual who had, in 1940, signed Vavilov's arrest warrant.

Now, as the German army rounded on Moscow, Beria ordered the transfer of all prisoners out of the capital. If these enemies of the people were to be shot, Soviet troops, not German soldiers, would pull the triggers. Captives from the Sukhanovo and Lefortovo Prisons had been the first to be moved, to an old prison in Orel, where they were executed. Then, at midnight on October 16, shortly after news of the German advance caused a mass panic among Moscow's citizens, guards collected Vavilov from his cell. They marched him and his fellow prisoners four miles to the Kurskaya Station, in the east. It was the first time in four months Vavilov had left the compound. But this was no walk to freedom. At Kurskaya the guards organized the men into rows and ordered them to kneel.

At first light, Moscow's residents emerged from their homes to witness the eerie scene: thousands of prisoners, the outlines of their emaciated bodies visible through soaked clothes, huddled on the ground. Here were the "wreckers" and the "saboteurs" about whom the Russian people had heard so much, and among whose ranks they prayed they might never feature. Perhaps to create some sociopolitical distance between themselves and the captives, some onlookers jeered, "Spies! Traitors!"

After the prisoners spent six hours kneeling, guards ushered Vavilov and his cellmates into a *stolypin* train carriage used to transport convicts. As many as twenty-five men were packed into carriages built to house five. Vavilov and the others took turns to sit on the carriage floor. As the train wended across frozen landscapes, hours turned to days. Some passengers passed out through fatigue and lack of air.

Vavilov's train was destined for Orenburg, but German air raids

forced the driver to divert to Saratov, close to the border of Kazakhstan, where, more than two decades earlier, Vavilov had mounted some
of his first expeditions as a young, enterprising explorer. On October
29, 1941, two weeks after the train departed, Vavilov arrived almost a
thousand miles to the southeast of his beloved Leningrad, his Institute,
his staff, and his seeds, no longer the golden child of Soviet science, but
its banished reject.

II

THE PRISON IN SARATOV WAS a place of darkness and despair, a
sprawling complex of barbed-wire fences surrounding gray buildings.
The landscape was desolate, a frozen wasteland of snow that stretched
beyond the horizon. The wind howled through the barren trees, carrying with it the faint sounds of distant cries and moans. As the captives
disembarked from the cramped, reeking train, they had only a moment
to take in their new surroundings before, once again, guards ordered
the men onto all fours. Then they called each man in turn to stand, step
forward, and strip naked.

A guard escorted Vavilov, once searched and showered, to Cell
Block 3, an area of the prison used for the most high-profile political
and public figures. Night fell. From his cell Vavilov heard the intermittent jangle of keys and the buzzing of the light bulbs that flickered overhead, casting eerie shadows on the faces of the inmates. The corridors
echoed with groans and the dull thuds of beatings, as elsewhere officers
worked to extract confessions or information. In his cell, Vavilov had
time to dwell on the people and projects to which he had dedicated his
life, especially here in Saratov, a city that had given him his lover.

At the time of Vavilov's arrest, Yelena and Yuri had been living
close to the Plant Institute's experimental farm at Pushkin, outside
Leningrad, a few miles from Kameraz's fields of exotic potatoes. With
German troops advancing, and the precise whereabouts of her lover
unknown, Yelena had decided to take her son and flee farther inland.
She moved to a dacha owned by a friend of the family's outside Moscow. Yelena delivered food parcels to Lubyanka Prison in the capital,

believing this to be where Vavilov was held, but as it became clear the German troops were closing in, she again arranged to return to Saratov, her hometown, five hundred miles from the front. The house had a vegetable plot, and Yelena's cousins lived in the house next door. After arriving, thirteen-year-old Yuri enrolled at the No. 21 Saratov school for boys.

The nights were sufficiently cold that mother and son slept in their coats. Yelena suffered from a nerve condition that limited the movement of her hands, so Yuri chopped the wood to heat their stove. When Yelena applied for a role at the Faculty of Agronomy, the institute where she'd first met Vavilov in the winter of 1917, the administration told her there were no jobs for family members of an enemy of the people. Yuri became responsible for not only fuel, but also food; on weekends, when there was no school, he visited a local factory that produced sunflower oil and collected the residue of mashed seeds, which he and his mother chewed for a dribble of sustenance.

Vavilov and Yelena spent the end of 1941 a few miles from each other in Saratov, each believing the other to be living in Moscow. Undernourished, lonely in solitary confinement, and psychologically stricken by the pain of separation from family, the estrangement from his colleagues and his work, the loss of status and identity, and the oppressive clouds of uncertainty concerning his fate, Vavilov's health deteriorated. He developed scurvy and began to wonder whether he would ever see Leningrad again.

III

AFTER DINNER AT THE WOLF'S Lair on January 12, 1942, while Vavilov grew sicklier in Saratov, Hitler directed his rage toward the heavens. The Russian winter had conspired against his mission on Moscow. Temperatures had fallen so low that his aircraft were unable to fly, tanks could not start their engines, and even the firing mechanisms on his troops' machine guns had ceased to function. For several weeks the conflict in Russia had been all but frozen in time. The recent fall of the average temperature from -2°C to -38°C (28°F to -36°F) was, the Führer stormed at his audience, an "unforeseen . . . catastrophe," one that had "paralyzed everything." The battle for Moscow, it was now clear, had become Germany's first major defeat in the Second World War.

Hundreds of miles away on the outskirts of Moscow, one of Hitler's soldiers, the twenty-six-year-old artillery officer and plant breeder Heinz Brücher, had experienced firsthand the difficulties of fighting a war in tyrannical weather. A few months earlier Brücher had stood at Moscow's gates and surveyed its monuments through his binoculars. At that moment victory in the east had seemed assured.

Then the mud and the sleet and the snow had ushered in *rasputitsa*—a period in the Russian calendar when sludge makes roads unnavigable. Without the proper equipment the invaders, who were already wearied from their journey across Russia's daunting expanse, were left immobile and vulnerable. In mid-December Hitler sacked Fedor von Bock, commander of Army Group Center, and assumed control of High Command himself. Operation Typhoon had failed, and Brücher now faced the prospect of retreat.

Brücher was a bright, well-educated plant breeder, whose fervent support of the Nazi regime had inspired him to volunteer to take part in the invasion of the Netherlands and Belgium. He shot through the ranks to the position of lieutenant. Swept up in the fast, successful invasions of the early months of the war, Brücher felt the power of the Wehrmacht was unrivaled, unstoppable. Then: Moscow.

Eighteen miles outside the Soviet capital, Brücher's unit halted its advance. Faced with Zhukov's proactive resistance, they soon abandoned their winter quarters and retreated into the snow. The withdrawal was bleak and arduous. Brücher and his men traipsed with frostbitten feet through woods and across fields, past the bodies of their comrades who lay splayed like roadkill in the snow. Injured horses, their heads drooped in resignation, wandered about without riders. At night the men crawled along frozen ditches while the glow of burning villages soaked the sky. The cold, Brücher reported, was relentless and pervasive. The abiding sensations were, he wrote, of hunger and frostbite, a renewed and sobering sense of the human body's vulnerability to the extremes of the outdoors.

"There is nothing special or uplifting to report from here," he wrote in a letter addressed to a colleague at the University of Jena. "I do not wish anyone such a winter as I have now been through." Brücher remained confident, however, that the change of the seasons would bring with it a change in the German army's fortunes.

"Now all that is done," he wrote, "and we hope to be able to destroy the Russians for good this summer."

Brücher's hope for the domination of Russia was ideological; like his masters, he believed in the superiority of the German people, and the necessity of quashing the Communist ideal. He had, however, another reason to be hopeful of victory; a plan to seize, soon enough, a treasure of unfathomable value to the German people and bring it home with him. Brücher wanted Vavilov's collection.

IV

EIGHT YEARS EARLIER, WHEN HE turned eighteen years old, Brücher had enrolled at the University of Jena to study biology, zoology, and anthropology. The son of a vet, Brücher had spent his childhood years in nature exploring the forests close to his hometown of Erbach, in the mountainous region of Odenwald in western Germany. These youthful, outdoorsy experiences ignited in him a passion for plants and animals, which carried him into academia. The following year, Brücher transferred to the University of Tübingen, to study under the renowned plant geneticist Ernst Lehmann. Lehmann, who described Nazism as "politically applied biology," hoped to divorce the research of Jewish thinkers such as Albert Einstein from German science and to establish a field of Aryan physics that would underpin a "national biology."

Brücher's understanding of biology was warped by Nazi ideology, which Lehmann's work claimed could be applied to the sciences. Brücher soon viewed plant breeding and race theory as twin branches of genetics; research into one would illuminate understanding of the other. Principles of heredity applied equally to blood and soil. Brücher's scientific interests were usefully aligned to the goals of National Socialist politics. This keen, ambitious young botanist was well positioned to become a Third Reich prodigy.

Brücher's career plans were complicated by questions over his own background. His applications to join the Nazi Party were rejected, initially because of concerns he had Jewish heritage, then, once he provided evidence to the contrary, because of his involvement with the Freemasons, a group the Nazis believed were accomplices in the Jewish world conspiracy. Brücher wrote letters of appeal in vain.

His fortunes changed, however, while he was still a student. Lehmann claimed to have conducted experiments on willow herbs—*Epilobium*—that proved genes were present only in the nuclei of cells and not in their cytoplasm, an assertion that was contrary to the popular view. While working on his doctoral dissertation, Brücher repeated his supervisor's experiments and yielded different results. Lehmann, wanting to protect his reputation, tried to suppress his student's findings. Brücher used the incident to question Lehmann's political convictions, a bullish move that turned a dispute between a teacher and his student into a scandal, from which Brücher emerged victorious.

He was duly accepted into both the Nazi Party and Heinrich Himmler's Schutzstaffel, Hitler's elite guard force. Brücher began to contribute articles to the Party's monthly journal, *Nationalsozialistische Monatshefte*—which had for a time been edited by Hitler himself—writing about Ernst Haeckel, the scientist from whose work Nazi principles of eugenics drew.

Propelled by the winds of Nazi Party favor, Brücher returned to the University of Jena with his PhD and began to collaborate with Karl Astel, a senior SS officer and one of the architects of National Socialist racial policy. It was Astel who, in 1935, suggested that Himmler turn the University of Jena into an SS university. Both men shared much in common: a desire to improve the German race through breeding, and a belief that the poisons of tobacco and alcohol should be banned.

Brücher's fame and reputation in German science rose. By 1941 he was working with the Kaiser Wilhelm Society, Germany's most prestigious research institution, where researchers often received generous state funding. One superior wrote of the young geneticist:

> Considering his age [Brücher] has during and between the participation in major military campaigns . . . accomplished quite extraordinary scientific and organizational achievements, always showing his enthusiasm for action, willingness to make sacrifices, and impeccable soldierly behavior.

Brücher was well aware of Soviet plant science and believed that the seed samples that Vavilov and his teams had collected could, if seized, hasten Nazi plans to improve Germany's crops. The Russians had collected hundreds of thousands of valuable seeds and accumulated a vast bank of knowledge. Moreover, Vavilov had overseen the creation of an extensive network of experimental stations across the vast country, from the ice fields of the frozen arctic regions to the sunny banks of the Black Sea, some of which stations were now in German-held territory.

In these stations Vavilov's network of staff evaluated, refined, and adapted plant varieties to the local conditions. Brücher wanted Germany to adopt these methods, and to acquire the field stations' samples. He shared Vavilov's belief that the shrewd combination of varieties of plants with beneficial traits could produce new kinds of crops with improved resilience and yield. In this way, plant breeding promised to solve one of the greatest shortcomings of the wartime German economy: food supply.

"The conquest in the east has given us access to areas that will be of utmost importance for the nutritional needs of the German people in the future," Brücher wrote. When he returned from the failed assault on Moscow, he began to consider how he might secure the Russian seeds before it was too late.

V

IN JANUARY 1942, WHILE HITLER cursed the weather from the Wolf's Lair, a sixteen-year-old schoolgirl from Saratov startled as the window in her cell door clacked open. It was through this aperture that, typically, her meager meals would be passed. This time, however, the guard ordered her to collect her things. Irina Piotrovskaya had been arrested and imprisoned for the unlikely charge of "attempting to organize an attempt on the life of Comrade Stalin." For any prisoner branded an assassin—even a child—a summons like this could, she knew, signal the end.

The guard led Irina into the appalling cold outside. There she saw the outlines of bodies already standing in the yard, their faces close to the wall and hands clasped behind their backs. The guard shoved her forward, into a space beside a thin man in a black overcoat. Irina had endured a great deal, but now, faced with the prospect of execution, or whatever this was, the dam of her emotions broke. She began to sob.

"Why are you crying?" she heard a kindly voice ask.

She looked up and saw the man in the black overcoat had turned his face toward her. He was gaunt, with a little beard, and intelligent eyes. A delta of smile lines revealed his years and recent trials. Irina explained that she was frightened about where she was to be taken next, by whom, and about the nameless pains that now racked her body. She told him her name, and the charges made against her.

"Listen to me carefully," he said, taking two small steps toward her. "Since you will almost certainly survive this, try to remember my name. I am Vavilov. Nikolai Ivanovich, an academic. Don't cry and don't be afraid."

The man's words had a calming effect on Irina. It was the first time anyone had shown her kindness in months.

"We are being taken to the hospital," Vavilov continued. "They have decided to treat even me before they shoot me. I am being held alone in a death cell. Do not forget my name."

Vavilov began to tell the girl a story, about a group of Jews who were burdened by unwarranted taxes designed to impoverish them. While their victimizers took their money, the men began to cry. There was, however, no way to avoid paying, and over time the men lost everything via these financial impositions. Then something curious occurred.

When the men's finances had been completely ransacked and they no longer had any means to pay, their crying turned to laughter. Their enemies had succeeded in taking all their earthly belongings. But now there was nothing left to take, the men were finally free. Vavilov smiled at her.

As Irina was bundled into a Black Maria to be escorted to the hospital for treatment, she felt calmer than she had in a long time. And she pledged to do as instructed. She would not forget his name.

VI

BRÜCHER APPRECIATED THE INCREDIBLE POTENTIAL of plant breeding, not only to better provide for the German people, but also to improve his own chances and opportunities within Nazi science. A recent SS expedition to Tibet, led by the German zoologist Ernst Schäfer, had yielded hundreds of varieties of barley, wheat, and oats (as well as Tibetan dogs, cats, wolves, badgers, foxes, and other animals, all returned to Germany). By combining this material with seeds acquired from Vavilov's collection, Brücher believed he could lead the development of cold- and drought-resistant, high-yielding crops. The German army had abandoned its goal of taking Russia's capital, but other treasures, already won in the early months of war, were within reach.

While the richest bounty—the collection in Leningrad—still lay within the siege ring, advancing German forces had claimed territory that included several of Vavilov's network of field stations— plant breeding stations in the Ukrainian cities of Cherkasy, Kiev, Bila Tserkva, Khorol, Lubny, Uman, Kherson, Dzhankoi, Yalta, and many more. Brücher suspected these institutions held samples drawn from the main Leningrad collection.

Relieved to have escaped the hellish Moscow winter and returned to Germany, Brücher remained desperate to return to the Soviet Union at the earliest opportunity, not as a soldier but as a botanist, to save and salvage these specimens at the earliest opportunity. As soon as the prized city finally gave up, and German soldiers marched into St. Isaac's Square, Brücher would have free access to the greatest and most diverse array of plant material assembled in human history.

Could his fellow countrymen be trusted to oversee this material with appropriate care? The Nazis considered Slavic culture to be inferior. This ideological sense of disdain found expression in the destruction of cultural monuments. German forces ransacked the writer Anton Chekov's house and reportedly lit fires using Leo Tolstoy's manuscripts. German troops repurposed the Russian composer Pyotr Ilyich Tchaikovsky's house into a motorcycle garage. There was significant risk, then, that in the looting of a city, Leningrad's botanical treasures could be ruined, through the starting of careless fires or the rampaging of foot soldiers eager to take out their pent-up frustrations on the city's cultural and scientific objects.

More troubling still, if Brücher hoped to reach Leningrad to see for himself history's greatest collection of plants and seeds, it was not a vision that aligned with his party's leader. On April 5, 1942, Hitler, during one of his typically meandering lunchtime rants, was unequivocal about his plans for the city: "Since there is a question of the future of Leningrad, I reply that, for me, Leningrad is doomed to decay." If Hitler had his way, the city would be razed and its collection destroyed before any German scientist set foot within Leningrad.

Each month, Hitler's resolve stiffened toward the tenacious city, which he had hoped would be emptied of civilians the previous winter. "St Petersburg must be razed to the ground," he raged a few weeks later. The destruction had to be total, not only to eliminate an enduring symbol of progressive multiculturalism, but also, in Hitler's mind, to prevent any future regrouping of Soviet leaders. "I was furious when the air force were reluctant to attack the place from their bases in Kiev," he said. "One day it has got to be done, otherwise the Russians will return and try to set up a government there."

For now Brücher's plan to lead a unit into Soviet territory to claim Vavilov's treasures had to wait for the outcome of the political wrangling of his various superiors, and the fate of the city.

In the spring of 1942, staff at the seed bank received an official order listing who among them was to be evacuated from the city across the frozen Lake Ladoga. When they departed on open-back trucks, each scientist carried packets of seeds under their clothing tied to their bodies with string or material.

XI

THE ROAD OF GRACE

I

ON FEBRUARY 17, 1942, GRIGORI Aleksandrovich Rubtsov, senior researcher of the Plant Institute's department of fruit crops and one of the world's leading pear experts, hobbled onto the frozen concourse at Finland Station, the only one of Leningrad's five mainline railway stations still in operation. A few days earlier Rubtsov had opened his front door to be greeted by two colleagues: Nadya Katkova, a laboratory assistant at the department of cereals, and Galya Lebedeva, with the news.

The two women explained that he and several other Institute staff had been chosen for evacuation from Leningrad, along with around three hundred other survivors who worked for the All-Russian Academy of Agricultural Sciences. They would finally be able to rendezvous with the handful of colleagues who had escaped to Krasnoufimsk the previous autumn, before the siege ring had closed. Rubtsov had come to the station to catch a train to the shoreline of Lake Ladoga. There an awaiting truck would deliver him across the ice, and into the frozen Russian hinterlands, to begin the twelve-hundred-mile journey to Krasnoufimsk.

Lake Ladoga had been one of the last remaining, if meager, routes for food to reach the besieged city throughout the autumn. The last barge crossed the expanse—almost seven thousand square miles in area—on November 15, 1941, as the water began to freeze over. Then began a waiting game. Experts estimated the ice would need to grow to four inches thickness before it could sustain the weight of a horse and rider, seven inches for a horse pulling a sled, and eight for a two-ton truck loaded with food.

On November 19 the first transport regiment crossed the ice using a horse-drawn sleigh. Four days later a column of sixty trucks arrived at nightfall on the lake's western shore to attempt the first motorized crossing. It was a dangerous drive: the harsh and ever-changing weather conditions battered the unsheltered ice, limiting the carrying capacity of every van and truck. Every journey across the fissured, shifting ice was perilous. A surprise thaw at the end of November contributed to the loss of 157 trucks. Drivers soon learned to leave their cabin doors open, so they could leap to safety if their vehicle became stuck. Sinkholes appeared at random, which led drivers to shine their full-beamed headlamps across the white span, at the risk of drawing the attention of the Luftwaffe's bombers, from which there was no cover on the pristine wasteland.

Toward the end of December traffic began to flow more freely, after the ice had thickened to a depth of three feet. Tributaries of road carved out by tractors pulling snowplows wended across the lake—a major feat of engineering in the most challenging of circumstances. The ice road ran from the rail-loading depots on the Soviet side of the lake parallel to the German siege positions and provided a rat run along which food and supplies could make it into the city. But ineffective organization meant trucks delivered less than half of the two thousand tons of food Leningrad's leaders needed each day. The slightest disruption in the flow of supplies had an immediate adverse effect on the population, many of whom were in too advanced a state of starvation to be saved anyway.

Not until January 18, 1942, after the ice road had become a highway, did its traffic fulfill the necessary quota of supplies, and the authorities could begin the slow work of rebuilding Leningrad's stocks. The trucks moved in trailing convoys spread across six lanes—three headed in each direction. Drivers were supposed to maintain a distance of twenty feet between each vehicle, but the sense of urgency encouraged rashness, and most drove bumper to bumper. The introduction of the GAZ-AA truck, and the larger, three-ton ZIS-5, both of which could drive across the lake at speeds of up to forty miles an hour, cut transit times to around an hour, reducing exposure to enemy fire on the ice, and enabling the most determined drivers to make as many as three crossings a day.

Traffic guides, situated at five-hundred-yard intervals, directed the procession around gaps or rocks, which emerged or disappeared in the shifting ice throughout the day. Nurses deployed to medical stations on the shoreline provided care to anyone suffering from frostbite, helping to drain pus-filled toes. To dissuade German attacks, the Red Army positioned antiaircraft guns near the edge of the lake, while Soviet planes swept back and forth above the convoys as deterrents to ward off bombers. Still, raids were commonplace: up to eight a day. Fearing German infantry troops might attempt to attack the convoys from the ground, groups of motorized sleds, each carrying three or four Red Army soldiers, fanned out across the ice, watchful for enemy incursions.

Despite the route's success as a highway for supplies, the plan to evacuate citizens drawn up on December 6, 1941, stalled, causing a delay that precipitated the deaths of tens of thousands of people. Before Moscow's State Defense Committee ordered the evacuation of half a million citizens from Leningrad on January 22, 1942, only thirty-six thousand escaped via the ice road. Finally, by early February, a robust system was in place. The cargo would flow in both directions across the ice: supplies into the city, people out of it, a mass evacuation that would prioritize women, children, and the elderly.

Not everyone received the news gladly. Despite rising despair and the full collapse of people's faith in the Soviet system ("It would be better if our country became a German colony," concluded one desperate inhabitant), some were loath to give up their homes or bedridden relatives. Some feared that if they moved out of their apartments, others would move in, and their homes would be lost forever. Others were unwilling to abandon their work.

Throughout February and March, Director Orbeli pleaded for his staff at the Hermitage to flee across the ice. The building now resembled nothing so much as a gilded cave, its high, ornate ceilings cold and damp, its cellars—where more than forty corpses still lay awaiting collection—full of the stench of mold and decay (in midwinter the Hermitage's water pipes and sewerage system had frozen). In each meeting with staff members, Orbeli would lay out the reasons why it was of utmost importance to evacuate from Leningrad at the first opportunity. Doing so, he argued, would ease the defense effort and free up rations for those unable to leave, and for the museum it could save the lives of scholars and all their knowledge and expertise.

Their evacuation was, he argued, "of great national importance." One by one these experts handed Orbeli written statements explaining why they did not wish to leave the city, the museum, or, as they saw it, their posts. "I do not want to evacuate from Leningrad," read one refusal. "I will make every effort to restore my former strength and dedicate it to my dear Hermitage."

Many Leningraders, however, were eager to leave. Crowds clustered around evacuation centers, and there were wails of indignation at any superior who jumped the line. Some of those unable to secure a place or unwilling to wait for one to come up, chose to attempt the journey on foot, against regulations. When they arrived at the lake's shoreline, those who were not picked up and arrested by NKVD agents struck out onto the ice. This already perilous journey had been made riskier still by German planes, which had reportedly dropped clusters of mines onto the ice. The explosives were disguised as cans of food.

II

ORDER 25 ARRIVED AT THE seed bank on February 14, 1942. The directive, which was signed by the Soviet Agriculture Ministry and the

THE ROAD OF GRACE   171

president of the Academy of Agricultural Sciences, provided a list of botanists to be evacuated, and the names of those to remain in situ. Anyone who refused to be evacuated would, the document stated, be made redundant. Those too sick to travel would have to leave as soon as they were well enough. As well as the scientists, four of the seed bank's accountants would have to leave for Krasnoufimsk as soon as they had completed the Institute's annual accounts.

The orders brought a renewed sense of focus to those who had survived the winter, but also cause for grief. The chief librarian, Georgi Heintz; his deputy, Elizaveta Voyko; and the head archivist Anisiya Malgina were among those instructed to remain at their posts in Leningrad. All three were already dead, the news having failed to reach the relevant authorities before they compiled the list of names. All evacuees were ordered to fill any space in their baggage with seeds.

It had been an act of tremendous, life-threatening will for Rubtsov to collect the requisite stamps and paperwork required for the day's journey. Having saved up food from his negligible daily rations, that day he and his colleagues had trudged across the Liteiniy Bridge with heavy bags. Every man and woman walked on swollen feet, a painful symptom of edema, caused by malnutrition, where low protein levels in the blood cause fluid to be retained, especially in the feet, ankles, and lower legs. Outside the station the scientists, some of whom were wrapped in blankets, underwent a rudimentary health check—a laughable inspection, considering the circumstances of the past few months, when a deficit of nutrition had aged and reduced all but the most powerful or corrupt. Once checked over, each traveler was handed meal coupons, redeemable at their destination.

Rubtsov, like the others, carried packets of seeds under his clothing tied to his body with string or material. Each scientist knew that the risks to the collection came not only from the invading German army, but also the ignorance of the state. After one of Vavilov's students, Evdokia Nikolaenko, was arrested and sent to the gulag, the guards had taken the seeds she had been carrying, scattered them onto the ground, and trampled them into the dirt. If the botanists could not save the seeds, nobody would.

On the station concourse Rubtsov shuffled up to Ivanov, who had come to bid his colleagues goodbye.

"Don't think ill of me," Rubtsov implored, feeling guilt at having been chosen to leave. "I know I am a lousy assistant, but over there I might have a chance to pull through."

Through months of shelling, Rubtsov, an erudite and well-traveled graduate of two universities, who had identified several new species of fruit plants in Central Asia under Vavilov's guidance, had served as a firefighter. His burst of self-deprecation was unwarranted. Malnutrition had reduced Ivanov's broad shoulders, but not his compassion. He disregarded his colleague's apologetic farewell.

"You will make it to the other side of the lake," Ivanov reassured his friend and colleague. "And then everything will be fine."

III

THE TRAIN SOUNDED ITS HORN and heaved out of Finland Station. While Ivanov returned to his post in the seed bank, Rubtsov and his colleagues underwent an imperceptible reclassification. These men were no longer *blokadniki*, the women no longer *blokadnitsy*; they were now evacuees, defined not by their confinement, but by their emergence.

Before Rubtsov could witness the miracle of the ice road wending across Lake Ladoga's featureless expanse, he and his wraithlike colleagues had to first survive the journey to the shoreline. The distance was only twenty-eight miles, but it presented an unexpected test of endurance for the enfeebled passengers. The train moved slowly, providing a tour of the city's districts and sights—the square at Znamenskaya; the picturesque Tavrichesky Garden, once owned by Catherine the Great's lover, Grigori Potemkin; the intricate frontage of Herzen University; the fairy-tale cathedral at Smolny. Each landmark triggered different memories and emotions in its viewers.

Whenever the German artillery began to fire, the train paused on the track for what felt like interminable periods. The passengers had time to survey and evaluate one another. Class, status, or fortune were no longer signaled merely by clothes or jewelry, but rather via the irrefutable, unconcealable condition of one's face and body. Those who had maintained a calorific diet through cunning or connection appeared fulsome and healthy, with plump cheeks and strong limbs; everyone else peered through sunken eyes above hollow cheeks.

"Although I had little interest in those around me," one young woman wrote in her diary of the train ride to Lake Ladoga, six days before the scientists, "I could only notice the families of hospital employees look[ing] wonderful. In our carriage there was a fat little boy, the son of an official. Next to my [boy], he seems especially rosy and healthy, while my son looks deathly." A journalist accompanied evacuees on one of the first journeys out of the city. He overheard an official in their group boast, "During the blockade I never went hungry." Elena Kochina traveled with the son of a high-ranking food-supply official who bragged that his family had eaten "better than before the war; we had everything."

The seed bank staff huddled together. A wood-fired stove provided some heat in the carriage. A journey typically measured in hours took more than two days. The evacuees slept where they sat. Eventually the train stopped at Borisova Griva Railway Station, several miles from the frozen shoreline where the trucks to the mainland awaited. Rubtsov and his colleagues completed the first leg of their journey on foot.

At the shoreline people huddled around bonfires. The great feats of organization and resilience that had established the ice road across the lake were not to be found in the care and administration of the streams of evacuees embarking on a journey that, even at full fitness and strength, would have been demanding. Most nongovernmental evacuees discovered there were no allotted places on the trucks. You found a place wherever you could. Rubtsov watched tired and angry passengers vent at whoever appeared to be in charge. Swift passage, some discovered, could only be secured through bribes slipped to the drivers. Desperate people bartered bottles of vodka, packets of cigarettes, wads of money, wristwatches, and jewelry for a valuable space.

As government-affiliated evacuees, Rubtsov and his colleagues found it easier to embark on the next leg of the journey. The journey from Borisova Griva on the lake's western shoreline to Kobona on the east is fifty miles by the road that borders the southern shore; by crossing the ice, the journey was more than halved. It remained, however, perilous.

The skies presented the twin risks of bombardment by enemy planes or swirling blizzards. At night, the stars and beams of the ve-hicles' headlamps provided the only light by which drivers could see; the risk of colliding with a boulder or tipping into a watery break in the ice was severe. Then there was the risk of exposure. Most evacuees from Leningrad traveled in the backs of canvas-topped trucks, with little protection from the elements. Some individuals were too weak to hold on to the vehicle, and as the truck bounced over the ice, their bodies were flung over the side. Drivers would rarely stop to retrieve lost cargo. A woman soldier was even assigned to the route to gather up the corpses of any babies and small children flung from their moth-ers' arms the previous night.

The driver motioned Rubtsov and his colleagues into the rear of his truck and told the children to sit in the middle of the huddle. The scientists wrapped themselves in blankets while the driver fixed the tarpaulin covering over them and started the engine. As the vehicle chugged onto the ice, the scientists joggled in the back, pressing their bodies against one another for warmth and stability, while the wind ripped at the sheet above their heads. Rubtsov waggled his numb toes inside his boots to aid blood circulation, but felt nothing.

Midway along their journey, Rubtsov heard the thrum of a plane's engine. He pulled back a corner of the tarpaulin and spotted the shadow of an approaching German Junkers in the sky. Beneath the sound of the wind, the botanists heard an explosion sound some-where close by. Then, as the enemy aircraft banked around to line up a second attack, Soviet fighters arrived to drive the enemy away.

The truck accelerated and passed an overturned vehicle engulfed in flames and black smoke.

The truck arrived at the lake's eastern shoreline in the early evening. As the vehicle pulled up at the first of the huts near the hamlet of Kobona, nurses and orderlies ran across the snow to greet and tend to the frozen evacuees. Each passenger was helped off the vehicle in turn, at which point a nurse checked the person over, rubbing snow into his or her face and hands. One man seated close to Rubtsov was slumped immovable in the back; he had not survived the crossing. A nurse opened the dead man's coat and removed his papers. Rubtsov was alive, but unable to stand. The orderlies lifted him from the rear of the truck and placed him on a stretcher. Then they carried him to a nearby field hospital.

He was warm, but Rubtsov could not stop shivering. A nurse brought him a bowl of food, every evacuee's long-promised, long-imagined first hot meal. Rubtsov managed only two spoonfuls, then turned on his side. This, the first meal of freedom, turned out to be his last. Within a few hours, Rubtsov was dead.* As the nurse unbuttoned his shirt, she found the seed package tied to his chest. It contained four pounds of grain.

"Why?" the young woman cried in disbelief. "He might have saved himself."

Another of the seed bank's staff, the fruit expert Pyotr Bogushevsky, lay in a nearby bunk with frostbitten face and feet. He raised himself onto one elbow.

"For life," Bogushevsky replied to the nurse. "For the children you'll have after the war. He was carrying seeds. He was keeping them warm. Give them to one of our people. Whoever makes it out alive can hand them on."

IV

LEAVING RUBTSOV'S BODY BEHIND, HIS colleagues continued by road from Kobona to Voybokalo railway station, a seventeen-mile drive

---

* For details about the precise date of Rubtsov's death, see note on p. 342.

from the lake's eastern shoreline. The hardship of the drive across the lake was nothing as to the ordeal that awaited.

At Voybokalo impatient crowds waited to board the breezy train wagons—*teplushka*, rudimentary carriages lined with bunks—that would carry them into the Russian heartland. Every passenger had to elbow their way aboard, then further hustle for a spot as close to the hay-stoked *pechki* stoves as possible. For the seed bank staff, the journey to Krasnoufimsk was interminably long: twelve hundred miles, almost the full length of Great Britain and back, through wintery squalls in freezing temperatures. Each staff member focused on the warming thought that, in Krasnoufimsk, they would be offered beds, hot meals, and the chance to add their precious packets of seeds to the portion of the seed collection that had made it out of Leningrad by plane a few months earlier.

The journey, however, was characterized by indignities and proved to be, for many, fatal. The trains lacked sufficient fuel and enough locomotives to pull the heavily laden carriages. They traveled at low speeds, often covering less than a hundred miles a day, thereby increasing the time the passengers spent in the freezing conditions. Trains often became stuck in deep snow on the tracks, which could only be cleared by enlisting the help of local inhabitants. Some remained stuck for six days, with insufficient provisions to feed everyone aboard.

One sixty-five-year-old engineer spent five sleepless nights in his windy carriage, which he shared with a group of emaciated engineering students. When one of the students died, another pushed his corpse out the window. The carriages were without bathrooms, an issue for travelers suffering from acute stomach disorders brought on by starvation. Relieving oneself required the support and cooperation of one's fellow passengers, who would open the freight car door, then support the individual by the arms as the person lowered his or her trousers and stuck their rear out of the opening. Those who managed to wait could enjoy the luxury of disembarking to squat trackside. Locals, according to one survivor, would watch on in disbelief, but having

endured far worse horrors, the passengers from Leningrad were indifferent to such judgment.

An NKVD report, published on March 5, 1942, three weeks after
Vavilov's staff began their journey, described the treatment of evacuees
by colleagues at a reception point as "irresponsible and heartless," and
the condition on the *teplushka* as "inhuman." Death was as commonplace in the carriages as it had been in the city. Seventeen corpses were
removed from a train at Volkhov Station; twenty at Babayevo; seven
at Cherepovets; and seven more at Vologda. In total twenty thousand
people, mostly fleeing Leningraders, were buried in a mass grave at
Vologda station, almost four hundred miles from where their journey
had begun at the lake's edge.

Three seed bank staff members were among the dead. Another,
the wheat expert M. M. Yakubinster, was removed from the train to
die, only to recover and continue his journey to the east, a small bag of
wheat seeds tied to his chest.

V

WHILE THEIR COLLEAGUES RODE THE evacuation trains across the
Russian landscape, in Leningrad the Plant Institute felt cavernous and
forsaken. The number of scientists living on the premises had reduced
from more than thirty the previous autumn to just fourteen. Their legs
weakened and swollen with edema or other complications related to
malnourishment, most required the use of a walking stick. The rodents
were gone, but the risk of looting by desperate human survivors remained. While shelling from German long-range guns persisted, there
was no aerial bombardment for the first three months of 1942, so the
risks on Leningrad's streets were somewhat reduced. But anguish and
desperation reigned, even as the inflow of supplies began to increase
provisions.

In late January, Leningrad's leaders had ordered the Hermitage
Museum's director, Iosif Orbeli, to establish a convalescent center in
the building across from the Plant Institute. Hermitage staff would
tend to the institution's own sick and dying, as well as those stretch-

ered in from four other museums. Staff set up a hundred beds on the ground floor of de la Mothe's Pavilion, and several women were chosen from among the Hermitage's keepers and curators to tend to the sick. And while across the city thousands lay similarly sick and dying in their beds, never leaving their homes, some still had the strength to hunt for resources.

One night in April 1942, a few weeks after their colleagues left for the ice road, Ivanov and his coworkers awoke to discover someone had successfully broken into the building by squeezing through the boards securing the ground-floor windows. From the sealed crates left by the soldiers in the entrance hallway, the thief had taken several hundred bags of seeds.

Even if the skeleton staff had heard the boards and crates creaking open, there would have been little they could do to dissuade the intruder; certainly, no one had the strength to physically restrain a determined thief. And besides, Ivanov reasoned, most of the seeds that had been stolen were no longer fertile, now useful only for reference. To outsiders, Ivanov's argument sounded illogical. He had convinced his colleagues to abstain from eating the same seeds he now dismissed as being infertile. And yet, scientific rigor meant Ivanov would be unable to eat specimens even if he believed them to be unviable. For every nine duds there could be one still living. The staff reboarded the windows, nailing boards to both the inside and outside of the frames.

The swift decline in population, brought on by the first two months of mass death—December and January—combined with the opening of the ice road for evacuees, enabled the authorities to increase the bread ration. On January 24 it increased to fourteen ounces for factory workers and technicians, and nine ounces for adult dependants and children. Three weeks later, on February 11, the authorities again increased the ration, to eighteen ounces for workers, and eleven ounces for dependants. Then, on March 22, another four-ounce increase for most inhabitants.

Water supply failures meant that for several days at the end of January and beginning of February bakeries were unable to bake and distribute bread. Even when food reached the bony hands of Leningrad's remaining population, starvation's long-gathered momentum was not easily halted. Deaths continued to rise during the first months of 1942. Children were left without carers; parents rendered childless; the capacity for grief was squeezed from the people. Death had become routine, as ordinary and expected as a shadow at sunset.

While the Plant Institute's rooms sat quiet and empty, the work continued. With his thirteen remaining associates, Ivanov oversaw the preparation of a new batch of seeds, once again destined for the relative safety of the Urals. The order for this dispatch arrived the day after most Institute staff evacuated. This time the plan was for Yan Yanovich Virs, deputy director of the Institute, to escort the samples—including a selection of potatoes saved by Kameraz—to unoccupied Soviet territory by plane. The departure date, a few weeks after the evacuation of the scientific researchers, was set, and Ivanov and his colleagues began to prepare a selection of forty thousand seed packages, weighing half a ton, and a full duplicate set of potato accessions for the journey by air.

Nobody complained. Under the light of cold-blast kerosene lanterns and during round-the-clock shifts, the team filled the sacks in silence to conserve their strength. On the day of departure, Red Army soldiers arrived, once again, at the Plant Institute, this time to escort an alternate selection of seeds and potatoes from the premises, away from the threat of German invaders. Virs bade a grave farewell to his faithful, surviving colleagues. There was no guarantee that either party would survive the days that lay ahead.

VI

SHORTLY AFTER VIRS'S DEPARTURE, THE remaining staff received a telegram from Eichfeld, director of the seed bank, who had successfully reached Krasnoufimsk the previous autumn. Eichfeld, perhaps having now learned of the true extent of difficulties since his departure, provided the remaining staff with a justification to consume the seed collection. If their survival depended on it, those left behind could eat the seeds, just as their forebears had done during the 1921 famine. The telegram read, simply, "Spare nothing to support people."

The message was clear: nobody would begrudge the scientists if they chose to consume the remaining stocks of seeds, particularly those held within the duplicate collection. Spring, with its new shoots and revived hopes, was close; some of the collection could be spared to ensure the survival of the faithful few, so that, come the dawn, they might oversee the planting and help revive the city.

Despite the clarity of their leader's instruction, the remaining staff rejected the notion. A single copy of Vavilov's hard-won collection was insufficient, Ivanov argued. All duplicate samples had to be preserved. "The war will be over one day and that's when we'll be held to account," Ivanov told his colleagues. "They'll ask what right we had not to protect the collection."

Eat or abstain? A simple choice that masked a clutch of complex, even taboo questions: Is any sacrifice justifiable in the name of scientific progress, or to protect one's research? What responsibility did the botanists hold to the survival of future generations? What was the correct course of action when that responsibility plainly sat at odds with their obligation to the living? There is no doubt that the quarter of a million seeds, nuts, and vegetables in the building, some of which, by Ivanov's own admission, were probably no longer viable to germinate, could have prolonged the lives of the botanists and, beyond that, the public. Equally, every scientist in the building understood that the seed bank's purpose was in part to provide a buffer against famine caused by plague, pestilence, floods, or, most pressingly, war.

There was no energy to waste their foggy, reduced minds and shallow lungs on heated debate. But, even if the surviving scientists weighed their options together or brokered disagreements about the ethical course, none recorded these conflicts or later recalled them. Dissent was either a personal matter or a private debate. This much is certain: Kordon composed a response to Eichfeld that made plain the scientists' final choice: "All our efforts are being directed toward preserving the collection. Any other questions are of secondary importance."

The task represented an enormous challenge. As Ivanov put it, "The fate of the collection now depended on the endurance of this small group of emaciated people."

In the summer of 1942, Leningrad's authorities instituted a city-wide growing campaign. The area in front of the seed bank (*above*) was turned into one giant plot where growers planted cabbages.

XII

CITY OF GARDENS

I

THE SEED BANK HAD BECOME a tiny, cloistered, traumatized community of scientists and caretakers. Each member of Vavilov's team was either dead, malnourished, or wearied. All had been crushed, to one degree or another, under the combined weight of Moscow's malevolent bureaucracy, the relative indifference of Leningrad's leaders, and the inexorable hardships of siege. Yet on they went, pitifully doing their best.

Klavdiya Panteleyeva called a meeting in the Institute's duty room. Together, she explained, the surviving staff needed to decide how to proceed for the next few weeks if they were to survive the lingering effects of winter. Consuming the collection was unthinkable, but Ivanov had recently made a discovery that could save them.

During a routine lamplit inspection of one of the storerooms, Ivanov noticed a burlap sack lying in a dark corner. He assumed it contained rubbish that someone had swept up and then forgotten about. Ivanov put down the lamp and dragged the sack into the open to check. Inside he found it contained as much as eighty pounds of Abyssinian peas. Someone had left them there in haste. Ivanov called Kordon in from the office next door.

"Are you sure this is right?" asked Kordon, not quite believing the staff would have overlooked a sack. The seeds were whole and intact and, crucially, eating them would not endanger the main collection.

"Someone must've brought this sack in just before the war and, in all the confusion, just left it there," said Ivanov. "In those days such a thing was of minor importance."

To reduce the risk of scurvy, the botanists devised a plan to germinate vitamin-rich sprouts from this unexpected source.

Ivanov planted each pea in a layer of calcined sand spread across a metal tray. Then he covered the specimens in cloches. Only one room in the Institute was sufficiently warm to germinate the seeds: room no. 12, occupied by the on-duty overseer, located on the second floor of the Institute's building at 44 Herzen Street. Throughout the day and night, the staff fed the potbelly stove scraps of wood to maintain an even temperature of between 10°C and 16°C (50°F and 60°F). On the fifth day, the seeds began to sprout. By the eighth day, the staff began to harvest thirty pea sprouts per day, per person.

This life-sustaining gift of nutrition was insufficient to save everyone inside the building. For some of those in advanced starvation, the body had consumed too much of itself for those losses to be restored. Calorie deficiency had burned up the body's stores of carbohydrates and fat, then moved on to the cells and tissues of a person's muscles and vital organs. Even their brains had shrunk, making the placid tetchy, and the focused muddled. A sudden increase in calorie intake could prove fatal, a disorder today known as refeeding syndrome, causing cardiac, pulmonary, or neurological collapse.

On March 15, 1942, as the sun began to break through the wintery gloom and flowers blossomed, the Institute's accountant E. I. Dmitrieva died, having survived the worst of the winter cold and malnourishment. Seven days later, the researcher M. N. Lavrova succumbed to starvation. Finally, on April 14, the associate researcher Serafima Shchavinskaya became the siege's latest casualty in the building.

Despite these ongoing tragedies, spring brought with it the first rays of hope, the first sense in a long while that change could be coming. At a meeting, Kordon, his body wasted but his mind still active, restated the importance of maintaining a sense of purpose.

"We've held out this long because we've been working," he said.

"That's how we must continue. If we lose the collection, we'll be left with nothing to show for [our sacrifice]."

"We can't check all the storerooms every day," said Nadya Katkova. "There are forty of them, and only a handful of us."

The remaining staff devised a new plan: to consolidate the collection once again, into fewer rooms.

"Fine by me," replied Panteleyeva. "We'll relocate the ground-floor collection away from the cold and from burglars and make an inventory, so we know where everything is."

These words, easily spoken, were not easily implemented. None of the surviving staff could walk without the aid of a stick. Their bodies' stores of fat were depleted, and in many cases the organ damage was beyond the body's capacity for repair, even with adequate nourishment.

Three miles away from the seed bank, at the F. F. Erisman Hospital, pathologists who conducted postmortems on those who had died from "dystrophy" found that the livers of the dead had on average shrunk to two-thirds of their typical size and weight, while the mass of a human heart had reduced by up to a half.

"Having used up its supplies of fat, the body starts to destroy its own cells, like a ship that, having run out of fuel, is broken up to feed its own boilers," the hospital's fifty-four-year-old chief pathologist wrote. "We knew all this in theory, but now we could see it with our own eyes, touch it with our hands, put it under the microscope."

With withered muscles and breathless lungs, the remaining Institute staff began the arduous work of consolidating, once again, the collection—an act of keeping busy, more than one of practical meaning, but an effort in which each man and woman hoped to find salvation.

Some staff members concentrated on the Institute's library. They dried out dampened books and placed them in cabinets with relevant labels. Others focused on consolidating the wheat collection. One person would slowly push each iron box along the floor. When it was in position, another employee would help lift it. Each container, which just a few months earlier the seed bank staff had bundled into groups of ten at a rate of four thousand a day, now posed a monumental challenge. As

Ivanov and a colleague carried a container up the Institute stairs to the first floor, Ivanov would regularly need to sit and take a break. After a few moments he would rise, slowly, and steady himself on the banister while he waited for his dizziness to pass, before continuing the work.

Once the wheat samples had been gathered, the team started on the rye, then the barley, then the millet. When the highest-priority grains were stored, Ivanov and his colleagues moved on to the rice, then the buckwheat, the peas, the beans. The staff recorded their work: "306 boxes of corn grain; 1,334 boxes flax; 1,582 boxes of beans (various)."

The thousands of containers provided a narrow focal point for men and women for whom the perimeter of life had shrunk to a single building, their community to just ten haggard but equally determined individuals. Whenever one staff member found a colleague slumped on the floor in a trance, he or she would help the person up with words of encouragement. By nightfall all the day's supply of energy had been exhausted.

On one occasion, Ivanov began to climb the stairs as evening fell, heading to the duty room for a rest and some food. His strength failed before he reached the top; he slumped down on a step and lost consciousness. He was roused by the sound of screaming. He opened his eyes to find his colleague Galya Lebedeva shaking him by the shoulders, her concerned face wet with tears.

"What's all the commotion? Has something happened?" he said.

"I thought you were gone."

"But it was just your imagination," Ivanov replied, trying to reassure her. "There's no need to panic."

Outside, the German bombardment continued. The duty manager diligently logged the hours of artillery shelling each day, which could sometimes last for six hours, unbroken. Inside the relative safety of the Institute, the work continued in its slow, aching way. Eventually, the last of the containers was stowed. The collection had been condensed to fourteen rooms. Kordon, Panteleyeva, and Ivanov together conducted an audit, drawing up a plan of the new storage layout; if they did not survive the siege, at least there would be a record of the names and whereabouts of the stored seeds.

The inventory taking had yielded some interesting findings. Some of the boxes contained old grain samples that the scientists believed had passed their expiry date and would no longer germinate. One collection of flax had been gathered eight years earlier, in 1934. For some reason these seeds had not been resown within a six-year period, per the Institute's standard procedure. Samples older than this were supposed to be removed from the collection. The discovery caused some consternation among the surviving staff.

"We're starving to death, and you're just letting them rot," cried one employee, upon the realization that her colleagues might have been spared by eating the dead seeds. "You're heartless. I want to live! To live! But all I hear is 'We mustn't. We don't have the right.'"

The senior staff met in Panteleyeva's office. Any supplement to the meager rations, someone suggested, could prove restorative and might prevent further deaths. With the expired samples the staff could make porridge in the mornings and soup in the evenings.

"I know it wouldn't be a violation," said Ivanov. "We can barely walk. And yet I can't agree to it. I just can't." Seeds age at different rates, he explained. The rule was clear: to store samples not for five or six years, but at least ten to twenty years. There could be a living seed among hundreds of dead ones, and that seed could germinate and produce an ear of wheat; some seeds had been known to sprout even after a decade's storage. "And we would have eaten it," he continued. "So, I propose we don't touch anything."

This faithfulness to the letter of scientific protocol was no longer an academic principle, easily pledged in times of peace and plenty. Ivanov's stubbornness, undeniably virtuous, had become deadly. And yet, for Ivanov, to eat even the expired seeds would betray the sacrifice of those colleagues who had died.

"That's the rule," Lekhnovich agreed. "We'll survive somehow till spring."

II

IN LATE MARCH 1942, THE temperatures after Leningrad's dread winter finally rose above the freezing point. Snow avalanched from roofs. Ice began to crack and dribble, exposing splintered beams, the bodies of the abandoned dead, and the tide of human excrement that encrusted everything, everywhere. Frozen pipes had left citizens to empty their slop buckets into any available drain, which by the advent of spring had become clogged and unusable. One onlooker recorded that the city was unrecognizable beneath the "piles of filth," loose garbage, and rubble that "cluttered up the courtyards and roads." Hercules's task to clean King Augeas's mythically filthy stables was, the onlooker wrote, "child's play" compared to the undertaking that now faced Leningrad's cleaners.

The authorities feared outbreaks of diseases could further annihilate the vulnerable inhabitants. Overstuffed mass graves began to thaw and stink. Without running water, unwashed dishes and tableware caked in fungal residue piled high in cafeterias across the city; meals were now served in the tins they arrived in, gobbled down with dirty fingers. The people were as grimy as their crockery: no baths, showers, or laundries had run since the end of December. To add to the risk of death by artillery, death by fire, death by cold, and death by starvation, Leningraders now faced a new peril: death by epidemic. Fear is always a matter of selection. For the beleaguered residents, long accustomed to the capricious risks of artillery fire, collective anxiety now fell upon the microscopic threat.

The mayor established an epidemiology committee, which drew up a program to vaccinate Leningrad's population against typhus, typhoid, and plague. In preparation for an epidemic, volunteers set up two thousand beds in children's homes. A cleanup operation was desperately needed to restore sanitation and a sense of civic pride. Volunteers posted leaflets on buildings across Leningrad urging citizens to clear their flats, rooms, and stairways of dirt and rubbish. "Let's protect ourselves and our neighbors from epidemics and save the lives and health of thousands of workers!" read one. After some false starts, the city council ordered all able-bodied Leningraders to gather in the streets on March 28, for the first day of the mass cleanup.

The turnout disappointed. Thousands arrived late or not at all.

There was insufficient transport, and too few crowbars and handle-less shovels available to equip those who came. Resentment toward the city's leaders, or the Soviet regime more generally, was rife.

"Let them feed us first, and then we'll work," declared one house-wife. "I don't intend to work for the Soviet government," stated another man, whose details were duly passed to the NKVD by an opportunistic eavesdropper.

Two days later, however, more than quarter of a million Lenin-graders turned out to participate in the cleanup operation.

"It's like putting a soiled north pole in order," recorded one writer and eyewitness. "Everything's a mess—blocks of ice, frozen hills of dirt, stalactites of sewage . . . When we see a stretch of clean pavement, we are moved. To us, it's as beautiful as a flower-strewn glade."

For the first time in months, crowds of people gathered. Another young woman who was sent to clear Tolstoy Square noted that people who did not have the strength to join in the cleanup, which resem-bled the "excavation of some ancient city," simply sat on stools in the sun, happy to be a part of the day's activity. Using sunlight narrowed through the lens of a magnifying glass, some onlookers lit cigarettes hand-rolled from torn scraps of newspaper.

Clearing a million tons of filth was disagreeable work.

"The stench from the half-melted chocolate snow is disgusting," recorded one able-bodied man, summoned to help clear the grounds of the Hermitage, close to the seed bank. "When you crack it with a pick, thousands of droplets splash onto your clothes and face." Workers rolled boulders of dirty snow onto plywood sheets, then pulled these makeshift sleighs to the banks of the Neva, where the cargo could be rolled into the water and washed away.

Close by, the walls of St. Isaac's Cathedral appeared to weep, as meltwater flowed in eager streams and pooled on the marble floor around the base of its pillars. The basement flooded. On the first dry day staff members opened the building's heavy cast-iron doors to re-duce the moisture levels inside. Able-bodied men and women carried the cathedral's valuables outside. Banquettes on slender legs steamed in the sun; multicolor carpets, curtains, and tapestries hung on ropes tied between columns; gilded rococo couches glinted among the monolithic columns on the southern portico; staff removed paintings from their frames and dried them out on the pavement.

As well as revealing hidden treasures, the arrival of spring exposed hidden horrors. The melting snow uncovered corpses that had been left where they had fallen or abandoned by friends, family members, and neighbors too weak to move a dead weight. Dead bodies lay in the gardens of the Anichkov Palace, on the banks of the Fontanka, and in the vaults of the Alexandrinsky Theatre. In the Nikolsky Cathedral one body had been placed inside a coffin, but twenty-three other cadavers lay on the stones wrapped only in sheets. The tally of the dead increased, from 89,968 burials in March to 102,497 in April.

Leningrad's pipes and drains weren't the only infrastructure to have failed during the winter. Post and telegrams, many of which carried crucial messages announcing the deaths of relatives, or their safe arrival following the evacuation, had gone undelivered for months; the great hall of the central post office was snared up with more than a quarter of a million boxes of undelivered mail. One eyewitness recorded postbags stacked halfway to the ceiling, unsorted and languishing in the unheated building. The first mail began to clear on March 8, with the delivery of around sixty thousand items. In the absence of postal workers, children volunteered to deliver letters and parcels. One young girl recorded how she went to deliver a letter only to find everyone in the apartment dead. When she returned to the post office, the building had been locked; the letter was then consigned to a purgatory, impossible to deliver, impossible to return.

The trams that had sat idle on their tracks for weeks began to run again. On April 11, 1942, the mayor signed an order directing the Streetcar Administration to begin running a minimal service along five routes in the city, starting from 6:30 a.m. The directive caused panic at the administration's offices, but 116 cars set out on April 15 as instructed. The sight of public transport running in the spring sun had a revitalizing effect on witnesses. Spectators cried on the Nevsky

Prospekt, Leningrad's main street, as with the sharpness of the blind they heard tram bells, the rhythmic clatter of wheel on rail, and the leaping flash of electricity sparks.

"I rode the streetcar!" exclaimed one witness. "I couldn't believe it. It seemed like I hadn't been on a tram for ten years." Those unable to squeeze aboard, or who were just caught up in the delirious elation of the old made new, stumbled after the carriages, laughing and crying. The leading ballerina of the world-famous Kirov Ballet, Natalia Dudinskaya, watched the trams trundle past from her apartment window above the scene. She had moved into Vavilov's abandoned apartment on the Nevsky.

One German soldier later told his Red Army captors that the morning of April 15, 1942, was when he lost faith in Hitler's mission: the sound of streetcars on Leningrad's battered streets was an astounding act of defiance. And yet, the sights and sounds of commotion, of a metropolis revived, were tinged with the melancholy of recent events.

"The city again is lively," wrote one resident. "A Red Army unit, probably convalescents, came by with a band. So surprising, so strange, after Leningrad's quiet. Streetcars are moving, jammed with passengers. On Bolshoi Prospekt there is trade and exchange. Money will buy more than in winter." Then, a macabre kicker: "Many are selling clothing—of the dead."

III

AS THE VALUE OF A human life collapsed during the first winter of the siege, so grief became a remote experience, a postponed idea. When life slipped away so easily and when every atom of one's being was angled toward survival, there was little capacity to mourn. The endless and oppressive sensation of hunger evicted all other feelings. "I lay there without any will, any desire, just in empty space," the poet Olga Bergholz noted of the total numbness she felt.

Starvation had brought with it a deadening of the emotions, a curious incapacity to lament losses that would, in fully fed moments, invite agony and sorrow. Mothers noticed a curious sense of alienation from their experience, a surprise that, as they tidied away the limp bodies of their children, they did not cry. As the human body diminished, so did the vividness and color of the human experience, as if an invisible hand had turned up the brightness and contrast on the world, so that it became more difficult to discern the texture of reality, till only the slightest of outlines remained, a whitening of the passions, a constriction of the margins of emotion.

"My brain no longer reacts to anything," wrote sixteen-year-old schoolgirl Lena Mukhina. "I can't feel anything at all."

Now the ice began to melt and with it the numbness. "The great sufferings of winter have gradually been replaced by a constant nervous irritation," wrote Lydia Ginzburg. Grief, which had been indefinitely deferred, also broke the surface. "Facts crept slowly out from the dimness of memory into the light of rules of behavior, which were now gravitating back to the accepted norm," Ginzburg continued. "She wanted a sweet so much. Why did I eat that sweet? I needn't have done." While the survivors of the dread winter could recall the facts of their behavior, they were unable to summon the sensations that had caused them to behave in what they now perceived to be "cruel, dishonorable, humiliating acts."

"All this winter I have been living only in the present moment," Georgi Kniazev, director of the archives of the Academy of Sciences in Leningrad, recorded in his diary on May 12, 1942. His body was warm again; the scrubbed city had begun sputtering to functioning life. The historian could now reflect on what and who had been lost. Sixty-five of the two hundred residents in his building had perished during the winter months.

"What an immense range of possibilities there are in man: soaring flights and plunging falls; geniuses like cosmic stars of the greatest magnitude, and monsters, depraved wretches; those whose names are known to history, and billions of those obscure people, unknown to anyone, who could never be remembered by anyone since nothing is known about them, but who lived and who died."

IV

As Leningrad thawed, physically and emotionally, the first shoots of new growth appeared in the city's green spaces. Wafted from the Baltic Sea, humid air carried the optimistic scent of willow trees. The green leaves of nettles, yellow dandelion polls, coltsfoot—*Tussilago farfara*—and chamomile broke through. The symbolism was naïve but impactful: new life emerged defiantly from amid the rubble. As the ambient temperature rose, so, too, did the awareness of the besieged people that they had survived.

"The enemy wanted to kill the city," wrote Ginzburg. "But the city lived on, and I was an almost unconscious particle of its resistant life."

After months of hunger and scarcity, the sight of wild plants caused a wave of frenzied activity. Families carrying baskets and pairs of scissors began to comb Leningrad's gardens in search of cuttings they could make into soups or eat raw on the spot. The parks became so full of botanical scavengers that one intrepid student at Leningrad State University ventured onto a firing range in search of specimens—anything that might provide a squeeze of vitamin C to combat the scurvy of winter.

During the winter, Leningrad's citizens had been nourished by the thought of planting vegetable gardens in the coming spring, a chance, finally, to assume some personal responsibility for their household's food supply. On March 9, 1942, the *soviet*—city council—announced plans to facilitate just such a program. Every family could now apply for a 130-square-foot plot within public green spaces. People could register to obtain seedlings from the Green Construction Trust.

This was the moment that Vavilov's staff had tenaciously waited for. Institute staff were now called on to help identify patches of land that might be suitable for growing vegetables. With this information the city's leaders drew up a list of 650 vegetable plots, each to be tended by people living in the vicinity of the allotments, for which the government canceled all taxes and rent payments. Every spare piece of land was turned into a vegetable plot: unused land in public parks and gardens, green spaces around St. Isaac's Cathedral, flower beds around monuments, lawns, riverbanks.

The Marsovo Polye—Field of Mars—swampland that had been drained and converted into parkland in the eighteenth century, was converted into a vegetable garden. St. Isaac's Square was turned into one giant plot, covered in cabbages. At the Hermitage, staff removed the lilacs and honeysuckle of Catherine the Great's rooftop "hanging garden" and replaced the plants with carrots, beets, dill, and spinach. Courtyards and vacant plots were dug up and planted with lettuce, beetroot, tomatoes, and cucumbers. Excrement cleared by the city's women was used as fertilizer. On the steps of the Kazan Cathedral a copper samovar bubbled while women drank tea brewed with freshly harvested herbs.

Leningrad's Executive Committee then instructed Ivanov to examine the urban areas and advise on which plants were fit for human consumption, as well as to help oversee the government-endorsed growing of seedlings and the distribution of equipment—the newly manufactured tools and wheelbarrows. Ivanov, his strength having begun to return from the new diet of pea shoots and increased bread rations, rose early to head out for the day, first exploring the parks and gardens in the city center, then venturing farther on foot, to the suburbs, where local farmworkers advised him on where to find plentiful supplies of certain herbs. In the rural settlement of Agalatovo, twenty miles north of the Institute, locals told Ivanov that tea brewed using beggar-ticks (*Bidens frondosa*), and blue cow-wheat stalks (*Melampyrum pratense*) could be used as medicine to treat scurvy.

Wild plants provided a rich and plentiful source of nourishment: dandelion leaves, nettles, goosefoot (*Chenopodium album*), and ribwort. But Ivanov identified poisonous plants among the edible specimens that posed a serious risk; fifty-two people are recorded to have died from eating poisonous varieties. Despite the risks, foraging became an industrial enterprise. Anyone could bring harvested specimens to reception centers and exchange twenty-five grams (one ounce) of edible plants for a bread ticket. Wild mushrooms and berries could be swapped for underclothes, soap, thread, tobacco, and vodka.

The raw plant material was sent to the food factory, where it was washed, sorted, and transformed into a variety of dishes to be distributed among the canteens. Dandelions (*Taraxacum officinale*) were used in various creative ways. Their roots were slightly dried, then ground and used for coffee. The flowers were marinated and used as a replacement for capers. The leaves were combined with the wild turnip* (*Brassica rapa*), pennycress (*Thlaspi arvense*), and shepherd's purse (*Capsella bursa-pastoris*) and served as a salad.

Seed bank staff drew up a list of edible plants, most of which were small and insignificant, typically dismissed by gardeners as unwelcome weeds, to direct inexperienced foragers. These recommendations were added to a map to direct foragers where to harvest different varieties of wild edible plants. Watercolor illustrations, painted by Professor Pinevich,† accompanied the list. Copies were made and distributed among the plant hunters and chefs. This information was then relayed via posters, which provided an at-a-glance checklist of edible wild plants: young nettles (*Urtica dioica*), dandelions, burdocks (*Arctium minus*), goosefoot, rapeseed (*Brassica napus*), and sorrel *(Rumex acetosella)*.

Lekhnovich, who had protected the potato collection through the winter while caring for his wife at home, remembered that the juice of rib-wort leaves (*Plantago lanceolat*) could be used as an anti-inflammatory for wounds and its infusion used to treat bronchitis, pulmonary tuberculosis, and whooping cough. As he scoured Leningrad's verges and footpaths in search of edible plants, he recalled his mother's collection of medicinal leaves: elderberries, roots of beggar's button (*Arctium lappa*), yellow dock (*Rumex crispus*), hog bean, nettle stalks, shepherd's purse—all tied into bunches and hung from the ceiling, each one to treat a different ailment.

"Girls in dinky kerchiefs, old-time working women from whose faces the shadows of the hungry winter have gradually disappeared, men who have not yet joined the army, old graybeards—all of them carry spades and are talking about planting, about early-maturing varieties of potatoes, carrots, and cabbages," observed the poet Nikolai Tikhonov in June 1942.

---

* Not the well-known vegetable, but rather a brassica—a member of the cabbage family.
† Probably Professor Lidia Mitrofanova Pinevich, who worked at the Leningrad Agricultural Institute.

Vegetable plots are springing up in the city itself, on the boulevards, on the lawns, in tiny garden patches, in the empty spaces between the enormous windowless backs of the houses and beside the monuments in the public gardens. Everywhere the plots are beginning to show green. . . . The city realizes that apart from everything else it has got to do, it must now help itself to improve the food supplies. Kitchen gardens! Kitchen gardens are also a front.

Factories and other enterprises organized more than six hundred so-called subsidiary farms—*podsobnye khozyaistva*—the produce from which went to the enterprises' cafeterias, the city's general food supply, and to the farm laborers themselves. Altogether, more than a quarter of a million Leningraders planted more than seven square miles of private gardens. In every patch of viable soil grew cabbages, lettuce, tomatoes, cucumbers, green beans, carrots, radishes, and onions—fruits and vegetables that provided a vital supplemental role in the diet of the besieged population whose bodies were now slowly recuperating from the winter of mass death. According to the newspaper *Pravda*, twenty-five thousand tons of vegetables would be grown on individual allotments by the end of the year.

Now, at last, the few staff members who remained at the Plant Institute could begin to see how their tenacity and restraint might benefit those who had survived the cold months of that dread winter. The team at the seed bank divided into smaller groups. Nikolai Ivanov and his colleague Vasily Ivanov took the train to Peri, thirty miles north. There they assisted local farmers in the Oktyabrsky District, providing growing advice and practical aid to the local population growing plants and vegetables in local plots.

The previous year's posters, which threatened "Death to provocateurs!" and urged loyal citizens to "Expose whisperers and spies!," were now replaced with words of encouragement to nascent gardeners: "Fifteen-hundredths of a hectare will produce 800 kg of cabbage, 700 kg of beets, 120 kg of cucumbers, 130 kg of carrots, 340 kg of swedes, 50 kg of tomatoes, and 200 kg of other vegetables." This harvest could feed "an entire family for the whole year," urged the copywriter, words that carried the unspoken encouragement to avoid a repeat of the previous winter's hardship.

It was arguably the most popular program that the city's Communist leadership ever organized—unsurprisingly, considering that every man, woman, and child understood in the clearest terms possible the elemental link between food and survival. And yet, distrust abounded, fueled by the equally vivid memory of those who had robbed and killed a few months earlier in the name of survival.

"On the benches along the paths sat motionless figures of watchmen," recorded one eyewitness of the scene at the Admiralty Park, where the grass had been dug up and replaced with even rows of tiny vegetable plots. There they would sit until curfew, watching over their tiny patches and fragile seedlings. "People don't trust one another."

V

THE POTATOES STORED THROUGHOUT THE first siege winter in the Plant Institute's cellar, where Lekhnovich had warmed them through the long nights with the heat of a potbelly stove, had survived and were now beginning to sprout. If the rare species were not planted soon, however, they would die, and the rescue mission would be wasted.

A duplicate set of the potato collection had been evacuated by plane from the city. But without confirmation of its safe arrival, Lekhnovich refused to count on the scheme's success. Mail from the Urals arrived only sporadically. If something happened to the evacuated potatoes, Lekhnovich might not learn about it for months. The botanist began to walk the city, searching for suitable plots in allotments or greenhouses within the siege limits. He and his wife needed land to plant the potatoes, preferably a sizable plot that would not require the collection to be split among multiple locations.

One morning Lekhnovich spied a plot attached to the Vyborgsky flower nursery, next to the Krasny Vyborzhets aluminum factory. The patch of land enjoyed plenty of sun and moisture and was well suited for growing the most valuable potatoes in the collection. It was far enough from the front line that he believed it would not be unduly bothered by shelling. The botanist made an appointment to meet with P. F. Nikitin, the acting director of the nursery, to ask for permission to plant the potatoes there.

"What potatoes? We don't have any," replied Nikitin.

Lekhnovich explained the situation, how, in a cellar of a building on St. Isaac's Square, he and his wife had protected many varieties of seed potatoes from rats, robbers, and frost, and that these vegetables now needed to be planted out.

"Have you not touched a single tuber?" Nikitin said incredulously.

"These are specimens of scientific importance, the fruit of the labors of a whole institute."

Nikitin slapped his palm on the surface of his desk. "Okay, we'll give you a plot! We'll help you to plow it. The horses are sickly, but we'll manage."

Lekhnovich and Voskresenskaya, assisted by the lab technicians Chernyanskaya and Lebedeva, collected two tons of potato samples from their cellar hiding place. They split the collection between two sites: the nursery by the Krasny Vyborzhets aluminum factory, and the Lesnoye state farm, to the north of the city.

At the nursery, the director assigned them a schoolgirl helper, a thin but, as Lekhnovich described her, "serious and diligent" eleven-year-old named Nina Afanasyeva, whose father had been killed during the battle for Tikhvin in the winter of 1941. The farm's blacksmith made the group a planter that had to be carried by two people, but which cut the earth deep enough to plant potatoes. Nina would walk behind the botanist husband-and-wife team, placing seed potatoes into the holes the couple dug out. If the tuber was small, Olga instructed, Nina was to place two specimens into the hole. Lekhnovich and his team would dive for cover whenever a shell landed close to the plot.

On June 22, 1942, the team finished planting out the entire potato collection, barring Kameraz's hybrids, in two parallel rows across the field. After days of waiting, the first shoots began to rise from the earth, with splays of tiny leaves.

While the seed bank's potato collection revived, the city authorities distributed seed potatoes between the collective farms—a few hundred pounds each—the basis for a crop to feed the city later in the year. To assist with the plan, Lekhnovich and Voskresenskaya drew up instructions to be distributed to farms across the city on how to increase the yield. Each eye of a potato tuber contains no more than four sprouts, the germs of the future stalks. The husband-and-wife team now recommended farmers cut the sprouts off, leaving just the stump. By doing so, the potato would begin to branch more vigorously, increasing the number of seedlings generated.

For sharing their knowledge and expertise, the director of the Lesnoye state farm, A. T. Vorobyov, allowed the seed-bank employees to eat in the farm's canteen, where they feasted on linseed-cake porridge, stewed saltbush, cabbage seedlings, and turnip seeds, which day by day began to revitalize the botanists.

Back at the seed bank Panteleyeva and Kordon remained in the city center protecting the parts of the collection that had not been evacuated to Krasnoufimsk. Each day the Institute's building manager, Maria Belyaeva, sat watch under the gates in the building's unheated entrance to ward off would-be burglars.

VI

On May Day the shops on the Nevsky Prospekt dressed their windows with artificial fruit and vegetables. One man recorded seeing drunks in the street—a sight that, in other times, might have kicked up feelings of disgust and judgement, but which this year gave him a reassuring sense that the world was righting its course. When he walked past a courtyard, he heard a woman weeping. Her grief brought him an unexpected sense of reassurance. After months of witnessing impassive figures dragging corpses on sleds, emotion had begun to return to the people. "Tears are proof," he wrote, "that the situation in Leningrad is improving."

Ivanov, too, found that sights and sounds that would have once irritated him now provided a comforting sense of normality. One day, while out identifying edible wild plants, he passed a school where he witnessed two boys fighting, a sight that kindled in him a sense of deep joy and happiness; only children with food in their bellies could engage in such an indulgently wasteful activity.

The shells continued to fall, but life inside the city felt different. The Leningrad correspondent of the Tass news agency recorded the happy sight of women, often wearing old army overcoats or workers' boots, carrying bunches of flowers: marigolds, violets, and dandelions, branches of spruce or pine, or simply handfuls of green grass. Another eyewitness noted that the scent of the lime trees in the Botanical Gardens "deadens the smell of decay from the rubbish, which is not yet completely cleared." All allowed themselves to contemplate, for the first time in a long time, a future.

"Yesterday we had an amazing evening," wrote Olga Bergholz in June. Bergholz had broadcast her wistful, honest poems on Leningrad Radio while, during the winter, she watched her husband starve to death. "At great expense Yurka bought a huge bundle of birch branches. We brought them indoors and put them in a vase. The window was wide open, and you could see the great calm sky. A cool breeze wafted in, the city was silent, and the scent of birch so sweet that my whole life, my best days, seemed reborn in me."

On the bank of the Lebyazhy Canal stood a tree that had been struck by a German shell. Its trunk had been splintered and twisted in the blast, its branches stripped. The tree appeared dead. But in the first days of summer, new shoots bearing tiny leaves punched through the wreckage.

PART THREE

Nikolai Vavilov arrived at the prison in Saratov—the same city he left two decades earlier to develop the seed bank in Leningrad—in October 1941. He was initially sent to Cell Block 3, an area of the prison used for the most high-profile political and public figures.

XIII

BEAT, BEAT, AND ONCE AGAIN BEAT

I

EACH MORNING ANDREI IVANOVICH SUKHNO listened to the ghoulish screams and scuffles of confrontation that sounded from the cell across the corridor in the basement where condemned men at Saratov Prison were left to pale and wither. The screams always arrived at a similar time: shortly after a guard passed the men their morning's ration through the slit in the cell door. The prison had been nicknamed the Titanic by its prisoners: in this multistory jail, most were doomed. But theirs was not the diet of a luxury cruise. Men awaiting execution received three meager meals a day: two spoons of kasha—groats—for breakfast, a tin of soup containing rotten tomatoes and a piece of salt fish for lunch, and a final spoonful of kasha and a slice of black barley-flour bread for dinner.

Guards seldom bothered the men who awaited the death penalty. The months prior to their condemnation had been a swirl of violence and bouts of unconsciousness, as the interrogators upheld Stalin's notorious order to "beat, beat, and once again beat." When the sentence had been passed, however, the state withdrew its fist. Each man was left in the shadows to contemplate his fate. Everyone in the death cells had reason to grieve their impossible circumstances. After months of interrogation, and enfeebled by hunger, who had strength left to shriek? Sukhno wondered. The sounds coming from the cell made no sense.

With little to lose Sukhno asked a passing guard about the source of the daily commotion in the opposite cell. The guard explained that, among others, it contained the disgraced academic Nikolai Vavilov, who had been sent there following his discharge from the prison hospital, and Ivan Kapitonovich Luppol, the philosopher and fellow member of the Soviet Academy of Sciences.

"But why are they shouting?" asked Sukhno. The guard walked on without answering.

II

IRINA PIOTROVSKAYA, THE YOUNG WOMAN who had pledged to always remember the name of the kindly academic standing beside her against the wall of the Saratov Prison courtyard, heard the news on the prison radio from her cell: Vavilov was dead.

The prison was impossibly overcrowded; there was no room to even lie on one's back. Packed together, the men could only shift position at night if everyone in the line did the same. Temperatures could reach as high as 30°C (86°F); men would routinely pass out from the stale heat in both the main prison rooms and the basement cells where Vavilov was incarcerated. For the vulnerable the conditions could be deadly, and yet the news report on the prison radio issued in the spring of 1942, just as staff from the seed bank had begun to plant the green spaces of Leningrad, was mistaken. Vavilov was a reduced man, suffering from scurvy and regular bouts of dysentery—but he remained alive.

While women and able-bodied workers cleared the streets of Leningrad, and his friends and colleagues advised citizens which wild plants could be picked and eaten, Vavilov had been moved from the sick bay to the basement cell. The conditions at Saratov were inhumane. Prisoners condemned to death were not permitted to write letters to their families, nor to receive parcels. They were denied access to the prison baths and given no soap. There were no books or journals to provide interest or distraction.

Wearing a canvas sack into which holes had been cut for his head and arms, and slippers made from the bark of the lime tree, Vavilov would perch on the fixed bedstead, pestered by the buzz of a single light bulb. Still, after enduring several months of solitary confinement, and surviving a spell in the prison sick bay, he had been relieved to arrive at a shared cell. He became close with his cellmates, especially Luppol.

To pass the time and provide interest to minds at risk of atrophy, Vavilov continued to arrange the small group. He invited each captive to speak in turn, encouraging the men to continue whenever they began to falter, urging them to whisper whenever their voices sounded too loud, lest they attract the attention of a passing guard. When it was Vavilov's turn to speak, he would recount stories of his adventures exploring in distant lands, carefully describing the vistas he had seen, the people he had met, and the plants he had collected.

Sometimes, Vavilov's focus would land on his current predicament, and the circumstances that led to it. He permitted himself to speak about Lysenko, describing him as a "speculator in the world of science" and explaining why his unscientific experiments would ultimately damage Soviet science. Vavilov did not shy from discussing Stalin, either. He recounted how, in 1936, when Stalin spearheaded a national campaign to raise the Soviet Union's grain harvest to one hundred million tons, Vavilov had authored an article pointing out that, prior to the Revolution, Russian farmers had regularly produced more than double that figure from the same production areas. This glancing criticism had won Vavilov an invitation to the Kremlin, for a dressing down by Stalin's right-hand man, Vyacheslav Molotov. Vavilov recalled how, during this admonishment, he caught the smell of pipe smoke in the room; Stalin had entered through a side door.

"Vavilov," he interrupted. "Why do you have to have these empty dreams?"

To instill a sense of order and schedule to their monotonous existence, Vavilov and his cellmates—three or four men squeezed into a space that, during peacetime, was used for one—organized a sleeping schedule: two men would sleep side by side on the bed, while a third dozed at the table that was fixed to the cell wall. In lieu of exercise, all but one of the men would lean against the wall, feeling the condensation on the palms of their hands, while the last person took a few paces backward and forward, to stretch his legs.

A routine formed: breakfast, lectures, rest, lunch, lectures, supper, then sleep. The men had lost the gift of daylight but tried to match their existence to its estranged rhythms. Then the arrival of a new cellmate destroyed the comfort of their fragile routine.

III

EVEN THROUGH THE FOG OF war, Vavilov's disappearance had not gone unnoticed by international friends and colleagues, who had become increasingly concerned about his fate ever since he had been forbidden from attending a ceremony in Edinburgh where he was due to be formally elected president of the International Congress in 1939.

Vavilov enjoyed particularly close ties to the British scientific community. He had first worked in England after he graduated from the Moscow Agricultural Institute and published his first paper there, on the immunity of plants, in the *Journal of Genetics*. His supporters included various luminaries of British science—William Bateson, Alfred Daniel Hall, Reginald Punnett, Arthur William Hill, Vernon Herbert Blackman, Rowland Biffen, and Cyril Darlington—most of whom promoted his election to the Royal Horticultural Society in 1931.

It was these professional friendships that provided credence to the NKVD's fiction that Vavilov was working as a spy for England. And it was these friendships that hastened the rumors that Vavilov had, in the words of Sir Henry Dale, "somehow fallen out of favor" with the Kremlin and vanished.

In part to express solidarity with a geneticist with whom many of Britain's leading plant biologists had worked closely, and in part to pressure Stalin's government into revealing Vavilov's whereabouts, on

April 23, 1942, twenty scientists signed a proposal for Vavilov to be accepted as a foreign member of the Royal Society, Britain's prestigious academy of sciences. It was a straightforward ploy: to accept the membership the candidate had to sign the diploma. If Vavilov's signature appeared on the paper, the society would know that he was at least alive, if not well.

The society sent the form to Moscow, addressed to Vladimir Komarov, president of the Academy of Sciences of the USSR, and head of the Botanical Department of the University of Leningrad. The Soviet academy had been evacuated to Alma-Ata in Kazakhstan, however. Undeterred, a British embassy diplomat traveled to Alma-Ata with the letter in hand.

Upon receiving the letter, Komarov, either of his own volition or on instruction from one of his few superiors, located Vavilov's younger brother, the physicist Sergei, and requested that he sign the document using only the family name, Vavilov.

The ruse failed. Soon after the British diplomat received the signed paper, the embassy issued the cutting response:

"We expected the signature of Nikolai Vavilov, not Sergei."

IV

Once again Sukhno listened to the sound of altercation from the cell across the corridor shortly after a guard delivered the morning's meager rations. A recent arrival to the cell, a history teacher called Nesvitsky, joined Sukhno by the door.

"I was in that cell for a few days," said Nesvitsky, who had been condemned to death for, he claimed, having "depicted the pharaohs in an un-Marxist way." Nesvitsky explained that when he was staying with Vavilov and Luppol, the prison guards had escorted a violent man suffering from mental illness to join them.

After the delivery of each day's meal this "madman," as Nesvitsky described him, attacked his cellmates to steal their ration. Vavilov and Luppol tried to reason with this vulnerable man, who should have been consigned to a psychiatric ward, but were left to fend off his attacks.

The food was insufficient to sustain a human body long-term—but losing one's daily ration would only hasten decline and increase the health risks.

The wild cellmate did not shy from punching and biting, however, and would often win the battle for the bread, leaving the two academics further weakened from the exertion, further reduced by the torment within a torment.

V

In Leningrad the commonplace cruelties of the gulag system had been exacerbated by the siege. Before the ice on Lake Ladoga began to crack in the spring sun, evacuees crossing the Liteiniy Bridge toward Finland Station would have passed an imposing building known as the Big House—today the headquarters of the Federal Security Service and formerly those of its predecessors, the KGB and NKVD— and Kresty Prison, the redbrick penal institution to which those declared "enemies of the people" by the agents at the Big House were often sent.

The hardships experienced by the general population of Leningrad were horrifically compounded at Kresty. Between October 16, 1941, and February 2, 1942, one inmate responsible for removing dead bodies from the prison kept a tally of the corpses he oversaw: 1,853. Every day he helped remove between twenty-five and forty bodies.

"The insides of their clothes were covered in a moving crust of lice," he wrote. None carried any tags of identification or markings. "These people were anonymous, nobody noted anything down."

The inmate and his coworkers carried the bodies into the prison yard, where they were loaded onto trucks and taken away. On February 3, 1942, around the time Vavilov was placed in the death cell with Luppol, all the cell doors in the Kresty Prison in central Leningrad stood ajar. There was nobody left alive to imprison.

Kresty was not the only one of Leningrad's prisons to be emptied by starvation. According to official figures, deaths in prisons rose from zero in March 1941, three months before the German invasion, to 3,739 in January 1942. For each of the next four months, the tally of prison deaths remained more than two thousand. Not only were prisoners viewed as having the lowest-priority mouths to feed during the famine, but they were also put to work on the ice road, logging camps and farms, and munitions, chemicals, and cable-making factories. According to records, the scant ration of nine ounces per day was insufficient for those prisoners put to work, who became quickly exhausted and "unfit for work." For thousands of Leningrad's prisoners, many of whom had been arrested on the most spurious of charges, the last of their human capacity was claimed by the state that imprisoned them and used to fuel the Soviet war machine. They were murdered by a fatal combination of slander, negligence, and exploitation.

VI

AFTER GUARDS FINALLY REMOVED THE mentally ill man from the cell, Vavilov petitioned the prison governor to improve his living conditions and routine. He also wanted to know what would become of him? The reply provided little reassurance.

"If we receive a paper from Moscow telling us to shoot you, we will shoot you," explained the governor. "If they tell us to pardon you, we shall pardon you."

Vavilov's fate was out of the prison staff's hands. By the spring of 1942, however, his condition had grown so poor that the governor granted the botanist permission to write a letter to Lavrentiy Beria, the debauched and genocidal head of the NKVD and one of the few men with the power to condemn a person or issue a pardon.

On April 25, 1942, Vavilov addressed his letter to "deeply respected Lavrentiy Pavlovich Beria." It was carefully worded, determined but cowed. Russia's greatest botanist had been reduced to pleading for an entry-level position in a field of science that he had, during his lengthy career, helped redefine. It was an appeal not for justice, but for mercy:

I am fifty-four, with a vast experience and knowledge in the field of plant breeding, with a good command of the principal European languages, and I would be happy to devote myself entirely to the service of my country, even to die for it. Since I am quite strong physically and morally, I would be glad at this difficult time to be used to improve the country's defenses in my speciality as a plant breeder, increasing the output of plants, foodstuffs, and industrial crops, . . . I request and beg you to make my lot easier, to decide on my future, and to allow me to work in my special field even at the lowest level.

Vavilov was one of hundreds of thousands of state prisoners who wrote similar petitions to Beria, commissar of internal affairs. Few reached him. But Vavilov, as one of the higher-profile convicts, was an exception. His letter wended its way through the NKVD's postal system, to the central office of the State Security Organization, on Dzerzhinsky Square in Moscow, within two weeks of being sent.

Beria, who had signed the original warrant for Vavilov's arrest, had the power to countermand his death sentence. Of the tens of thousands of cases that passed across his desk, Vavilov's remained unusually clear in his mind. Ever since the botanist's sentencing, Beria had received an avalanche of correspondence from Dmitri Pryanishnikov, the pioneering biochemist and plant physiologist who had taught Vavilov and, in 1941, won the Stalin Prize—an award for which, extraordinarily, Pryanishnikov had nominated Vavilov.

The veteran biologist tirelessly petitioned on Vavilov's behalf. Pryanishnikov was sufficiently powerful or courageous that he felt no qualms in providing vociferous support for Vavilov, not only to Beria, but also in letters to Stalin. Pryanishnikov had a personal link to the head of the NKVD that may have emboldened his petitions: for a time, he had taught Beria's wife, and she now worked in his department.

The combination of Vavilov's humble letter, the correspondence from the Royal Society in Britain, and pressure from his influential ally Pryanishnikov encouraged Beria to reconsider the case. Despite the dossiers of denunciations, the botanist's expertise might be of use to the Soviet war effort after all. Beria considered a compromise, one that would spare Vavilov while allowing him to save some face: a transfer to a specialist gulag known as a *sharashka*—a laboratory behind barbed wire—where the scientific knowledge of accomplished prisoners could be drawn upon by the state that had condemned them.

On June 13, 1942, Beria's deputy sent a special letter to the chairman of the Military Collegium of the Soviet Supreme Court regarding the cases of Nikolai Vavilov and his friend and cellmate Ivan Luppol:

> In view of the fact that the men under sentence referred to above might be used on work of importance for the country's defense, the NKVD of the USSR petitions for the sentence of death to be commuted to detention in the corrective labor camps ... for a period of twenty years each.

Five days later, in England, the Royal Society elected Vavilov as a foreign member for his "research on the origin of cultivated plants," as well as his work assembling the seed-bank collection. The proposal, which was signed by twenty Royal Society members, praised Vavilov's expeditions for having "furnished material for understanding cytological and genetical investigations." In his remote prison cell, Vavilov was unaware of the accolade, or the efforts of the international community to discover his whereabouts and confirm his well-being.

The effect of Beria's letter, however, was immediately felt. Ten days later the Presidium of the Supreme Soviet overturned its verdict after less than a minute's deliberation. The news reached Vavilov two weeks later: his life would be spared. If he survived the war, the gulag system, and the murderous whims of his jailers, Vavilov would be free—albeit at the age of seventy-three.

On the reverse of the official document Vavilov wrote, with defiance and joy, "This decision was communicated to me on July 4, 1942."

A guard collected the botanist and his cellmate and escorted the pair from the basement room to the communal cells. The floor was overcrowded, and the food rations no improvement, but Vavilov now enjoyed a suite of basic benefits: the opportunity to exercise daily, wash, and visit the prison shop, which stocked onions and tobacco.

But while Vavilov's situation had improved, the effects of long-term malnutrition had left him severely weakened. When, eventually, he became unable to walk, the friends he had made in the cell 57 carried him outside for fresh air.

VII

TOWARD THE END OF THE summer, the military situation around Leningrad had reached a stalemate—a static siege in stark contrast to the ferocious fighting unfolding around the city of Stalingrad, a thousand miles to the south. There the battle had begun in August 1942 when the German Wehrmacht, under the command of General Friedrich Paulus, a First World War veteran who had helped plan Germany's invasion of Russia, launched a vast offensive to capture the city. Stalingrad was a crucial industrial city and a major transportation hub on the Volga River. Capturing the city would have allowed the Germans to control the flow of oil from the Caucasus region, severely disrupting Soviet supply lines, and to deliver a symbolic blow to the city named after Stalin. But the Soviets employed tactics that prioritized close-quarters combat, utilizing the streets that German bombers had turned to rubble as defensive positions. Within weeks the battle had become a grueling attritional struggle.

Leningrad's survivors, suspecting they may have lived through the worst of the siege, became obsessed with the plight of their sister city and its faraway struggle on the Volga, which cast the privations of their current situation in a certain relief.

"All the time we feel for Stalingrad," one resident wrote in her diary.

There had been changes at home, too. Lieutenant General Leonid Govorov had assumed leadership of the Red Army at the Leningrad Front. A taciturn but warmhearted strategist, Govorov brought a fresh perspective and renewed commitment to the Russian efforts to decisively break the siege ring. In the lead-up to the performance of the Symphony No. 7 in C Major, op. 60, written by Dmitri Shostakovich, in the Philharmonic Hall, Govorov allowed Red Army musicians to swell the orchestra's ranks. And during the concert, which was held on August 9, 1942—the anniversary of the date the Germans had once boasted they would capture the city—Govorov ordered radio transmitter dishes to be set up so Soviet troops could listen to the performance and to play through speakers aimed toward the German lines.

Next, Govorov devised a system of decoys intended to draw the enemy's fire from its camouflaged artillery emplacements to reveal their locations. During the night of October 30, 1942, one of his lieutenants repositioned a 152 mm artillery battery in the Pulkovo Heights. At dawn he readied his gunners, who aimed at Army Group North's headquarters at Gatchina, and the nearby airfield and railway station. The troops let off a series of volleys, then ran for cover. "It was as if we had ripped open a hornet's nest," the officer recalled.

Sixteen different enemy batteries returned fire, a hail of shells directed toward the gun position. A group of Junkers appeared minutes later and began dive-bombing in relays. The lieutenant and his men were unharmed, and crucially, the ploy had worked: the Germans had disclosed many of their artillery placements. The heavily publicized news of the unit's gallantry proved inspiring, as did the reports coming from Stalingrad of a Russian counterattack. When, in October, Nazi leadership transferred the brilliant commander Erich von Manstein from Leningrad to the southern front to help rescue troops trapped in Stalingrad, Germany's siege forces shifted to the defensive.

As Govorov finalized his plans to break the German blockade, the botanists at the seed bank prepared for a different sort of winter from the one in which they had lost so many colleagues and loved ones. Pipelines newly laid beneath Lake Ladoga had increased fuel and power supplies. In the staff room, the botanists rolled cigarettes and reminisced about the hardships of the previous winter, and their improved lot.

By the end of the year more than four and a half thousand Soviet guns and mortars had been massed on either side of the German corridor to the east of Leningrad. In preparation for Operation Iskra, as the counterattack was known, Govorov had overseen the construction of a special training ground in one of Leningrad's suburbs. It featured mock-ups of enemy strongholds, dugouts, trenches, and ramparts built from wood, peat, snow, and ice. Infantrymen and women were drilled on firing ranges. "By the beginning of January, the musicians in our artillery orchestra knew their scores," Govorov later recalled.

At 0930 on January 12, 1943, the botanists at the Plant Institute felt the roar and rumble of Leningrad's guns as the Red Army bombarded the exposed enemy positions. The assault lasted for two hours and twenty minutes. When the guns fell silent, Russian troops charged across the ice of the Neva, while newly built Soviet planes provided air cover. The intensity of the assault proved arresting.

"It was a nightmare," a German sergeant wrote. "The shells exploded precisely where our bunkers were located." After his company commander died, panic spread among the unit; some men stepped out of their frontline trenches, their hands raised above their heads in surrender.

Three days later a Soviet ski brigade crossed Lake Ladoga. The men outflanked the German defenses and broke into Shlisselburg on the east. On the morning of January 18, 1943, the two sides of the Soviet forces broke through the German lines to contact each other, creating a narrow passage of land between Nevskaya Dubrovka and Shlisselburg. The Germans still maintained positions a hundred miles long and four miles deep, looping from the southwest of the city to Novgorod in the north. But a vital link between Leningrad and the Russian mainland had been prized open: a six-mile-wide corridor that led directly into Leningrad.

At 2300 that day, radio announcers broadcast a communiqué across the city. It read simply, "The blockade of Leningrad has been broken." The scientists and caretakers at the Institute joined the crowds of survivors to celebrate in a city no longer subject to curfew. Music resounded; speeches were made; drinks toasted.

In her diary that night Olga Bergholz wrote, "The cursed circle is broken."

VIII

SIX DAYS AFTER THE SIEGE circle broke, a thousand miles to the southeast, Vavilov was returned to the prison hospital. After his sentence had been downgraded, the botanist had reasons to be glad despite his ailing health. The news that he would be transferred to an agricultural camp—perhaps Dolinsky, near Karaganda, where thousands of imprisoned agronomists, physiologists, and plant breeders had been sent to work—had provided the rumor of a purpose. He would remain a prisoner, but would find freedom in his botanical work, which might, in time, facilitate a reunion with his friends and colleagues at the seed bank. But summer came and went. Then autumn. Then the New Year. With each change of the season Vavilov's health further failed, locked into an unbreakable trajectory of decline.

When he arrived in ward 12 on January 24, 1943, Vavilov's body was thin and wasted, his skin pale and translucent. The sick bay was full of sick and dying men. In the bed beside him lay Professor Artsybashev, a specialist on southern flora and another resident of Leningrad. In the thirties Vavilov and Artsybashev had clashed professionally. Now in their shared vulnerability, both men were pleased to be close to a friendly—if barely recognizable—face. Extreme and prolonged lack of food had left both men withered.

A nurse examined Vavilov and recorded the following symptoms: "Exhaustion, skin and mucosa pale, edema on legs." Vavilov was dying in the same way more than a dozen of his colleagues at the Plant Institute had died a year earlier.

When, during his rounds of the prison, the governor arrived at his bed, Vavilov asked that he and Artsybashev be given a glass of rice water each. The governor's temper flared at the meager request. He barked, "What on earth are you asking for? There isn't enough rice for wounded soldiers at the front, and you think I'm going to feed it to enemies of the state?"

It would be Vavilov's final request. Two days later, on January 26, 1943, at seven o'clock in the morning, the doctors pronounced him dead. Four days later a forensic examiner, Zoya Rezayeva, conducted the postmortem. "Corpse of a man of about sixty-five, medium height, average build, badly undernourished, skin pale, subcutaneous cellular tissue absent," she wrote in her report.

The cause of Vavilov's death was obvious. And yet, to ratify its name in an official document risked impugning the medical staff who had overseen treatment, the governor responsible for the distribution of provisions in the prison, the Soviet officials responsible for the supply of foodstuffs to the Russian people, and Stalin himself.

Perhaps feeling the combined weight of dozens of powerful men over her shoulder, Rezayeva listed "bronchial pneumonia" as the cause of death. Vavilov would be denied even the dignity of an honest death certificate.

A few days later, after sundown, a prison employee collected Vavilov's body, along with those of a dozen or so other inmates, and hoisted it onto the sleigh he used to transport them across the snow. Each man left the mortuary naked, save for a metal identification name tag attached to his foot.

Grave digging was not an official part of Aleksei Novichkov's responsibilities at the prison, but he was glad for the extra income and, just as important, the measure of spirits he was given as a disinfectant for his hands, which he would save and drink instead.

Novichkov hauled the sled into the frozen dark, past the tilted headstones to a remote part of the cemetery where the ground had been intermittently dug up for common graves. When every corpse had been thrown into the hole, he covered the crumpled bodies with soil, then knocked a metal spike into the dirt to indicate the plot had been filled.

A thousand miles from Leningrad, his Institute, and the world's greatest collection of seeds ever assembled, Vavilov had died in the same way as many of the men and women he'd worked alongside and loved. He had been murdered through a combination of false accusations and the deprivation of the food that he had dedicated his life to providing others. An oak tree had fallen.

Nikolai Ivanov boarded a tram that took him as far as the Kirov factory, five miles south of the Plant Institute, to plant some of the Institute's seeds in a farm close to the frontline. The fields were close enough to the German positions that a sniper could follow through his sights a horse-drawn plow.

XIV

A FARM ON THE FRONT LINE

I

A LITTLE AFTER TEN IN the morning of February 7, 1943, twelve days after Vavilov's death and more than five hundred days after the siege of Leningrad began, train no. L-1208 drew into Finland Station and hiccuped to a standstill. Trains had arrived this way tens of thousands of times, but never with such ceremony. The station, ruined with cavities, its locomotive shed a tangle of bent girders and broken glass, was festooned with bunting and resounded with the sound of a brass band. Decorated with oak-leaf-wreathed portraits of Stalin and Molotov, the train pulled two passenger cars and a caravan of freight carriages. Kittens—animals long absent from the city—stared back at astonished onlookers from the windows. The animals, sent to take on the hordes of vermin that had attacked the Plant Institute's seed containers, had apparently broken free of their boxes during the journey and moved into the comfier seats.

The train, which wore a banner that read DEATH TO THE FASCIST GERMAN USURPERS, had made the inaugural journey along the twenty-mile temporary line that passed through the narrow corridor prized open by the Soviet troops and that connected Leningrad with the mainland. It ran courageously across a stilted pontoon bridge on the Neva River. Supplementing the ice and barge routes over Lake Ladoga, trains now began to regularly run the gauntlet of short-range artillery fire, to ensure that food, raw materials, and fuel could continue to reach those in need. Whenever a shell exploded on the line, workers raced to fix the tracks, hastened by an elemental determination that the privations of the previous winter would not be repeated. Food was not plentiful, but it was available, which, for Nikolai Ivanov, only made room for other concerns.

As the snow had thawed and winter's chill gave way to a luke-warm spring, Ivanov had grown increasingly anxious about the looming sowing season. The puncture of the siege ring had given the surviving botanists at the Plant Institute reasons to be glad. But while parts of the collection would keep their latent life for years (the tomato seeds, for example, could sit in a container for as many as twenty years and still germinate when planted), the viability of many of the seed bank's cereals and pulses could not be guaranteed. Even if the cereals had withstood the harsh winter conditions when storage protocols were difficult to maintain, the seeds could have aged more quickly than anticipated. There was a risk that, when pressed into soil and watered, the specimens would no longer quicken. If the botanists failed to renew the seed stocks soon, more parts of the collection could be lost.

Ivanov asked Klavdiya Panteleyeva if, together with junior re-search officer Praskovya Petrova, and Anna Andreyeva, he could leave the Institute for the Predportovy, a collective urban farm situated to the south renowned for its fertile land. There the scientists could plant the two hundred varieties of seeds Ivanov had identified as being at risk. From these seeds new plants would grow, flower, and seed, en-abling the botanists to renew the collection.

Panteleyeva was hesitant. The surrounding German army had been pushed back to the Sinyavino Heights but retained most of its positions around the city, an arc of fortifications and artillery positions that swept from Pulkovo, south of Leningrad, to the rail junction at Novgorod, a hundred-mile defense line. Soviet forces were exhausted from Operation Iskra. The farm was only a mile from the front line. Panteleyeva had already mourned so many colleagues. She could not stand to lose any more.

Ivanov was resolute. The risks, he insisted, were just as severe in the city center. So many had died from shelling that locals had begun to refer to the square in front of Finland Station as "the valley of death," and the Liteiniy Bridge as "the devil's bridge." Besides, he added, staff at the Institute had given their lives to preserve their work. To fail to save the collection at this stage would betray their sacrifice.

II

IN THE EARLY SPRING IVANOV boarded a tram that took him as far as the Kirov factory, five miles south of the Plant Institute. The city farm was close enough to the German positions that a sniper could follow through his sights the bump and churn of a horse-drawn plow. As the tram approached the familiar factory entrance checkpoint, the ticket inspector informed his passengers that this was the last stop; they could proceed no farther.

Ivanov disembarked and continued on foot. As he walked through the suburbs, he passed the signs of violent close-range bombing: ruined buildings; trees whose branches had been stripped of leaves and whose trunks had been stripped of bark; jagged wet craters that pocked the landscape. At the bridge barricade he showed his papers and obtained a pass before continuing his journey. Near the eighteenth-century Krasnenkoe Cemetery he reached another security checkpoint. Here the defensive fortifications were taller and more numerous. Czech hedgehogs and dragon's teeth—antitank emplacements that had become a familiar sight at the perimeter—provided a sobering reminder that he was now deep within the enemy's striking distance.

On Stachek Avenue, Ivanov reported to the farm's offices, a huddle of buildings that included a canteen and accommodations for the farmworkers. Inside he met Nikolai Ivanovich Khlopkov, who had managed the farm since the previous spring, having been called back from the front. Khlopkov had gathered volunteers to work as vegetable farmers, mostly women who worked at the tram depot, and organized the planting and gathering of crops under constant shelling and bombing from the enemy positions a mile away.

Leningrad's people had relied on collective farms such as this to the north and south long before the siege began. Potatoes, cabbages, carrots, beetroots, and other vegetables were grown on land allocated to farming, to help maintain food supplies in times of need. Even before the Bolshevik Revolution and the man-made famines that followed, roughly one harvest in three was insufficient to feed the nation. These smaller suburban farms helped ease shortages, a role that had now shifted to supplementing the rations coming via Lake Ladoga.

Khlopkov introduced Ivanov to Ivan Vasilyevich Golikov, the farm's chief gardener, who was to be Ivanov's main point of contact. Golikov was taciturn and focused. He asked few questions, instead motioning for Ivanov to follow him to the planting fields. The pair arrived at a curious scene: a field crisscrossed with trenches and the dark scar of an antitank ditch. While Ivanov took in the scene, Golikov motioned toward a series of pillboxes in the distance.

"Those are German positions, right there," said Golikov. "Don't wander. They'll shoot."

"Could we plow that wedge over there?" Ivanov pointed at the curve of a trench.

Golikov laughed. "It's in full view." Golikov waved his hand at a heap of metal and rubber, what used to be the farm's tractor. "And even if you could do so without being shot, you'd have to do it by hand. We'll find you some old horse, for sure, but you won't be able to do much plowing with her."

Ivanov was undeterred. He and Petrova would plant at night, he explained. "We'll use spades. The nights are long, and if they start shooting, we can always shelter in a dugout."

Golikov nodded. He was used to finding creative solutions here at the extremity of human experience. The previous spring, after the dread winter, he and his helpers had plowed the fields and sowed seeds in the farm's greenhouses while barely able to walk. They had carried fertilizer on the same sleds used to transport the dead, pulled by three or four women wearing harnesses.

Each evening that week, after darkness fell, Ivanov and his colleagues stole along the trench, then clambered over the sides into the field. Positioned at short distances from one other, the botanists overturned the soil. Ivanov thrust his spade into the soil, turned the clump of earth upside down, and broke it into pieces, ready to be plowed the next day.

Ivanov's shirt soon became drenched with sweat. He couldn't work continuously without a break, a lingering reminder of the physical cost of the deprivation he had experienced during the past eighteen months. As the wind blew and the muted conversation of fellow workers drifted toward them, the group was keenly aware of the danger on the other side of the front line. From time to time an enemy shell tore into the night sky, leaving burning firework trails as it arced toward the city center.

One moonless night Ivanov's spade struck a rock, and the sound of clanking iron ricocheted across the field. A searchlight ignited from the enemy's position; its beam began to rove the bushes.

"Down," shouted Ivanov. "Get down."

Before she could drop to the soil, the light caught the Institute's caretaker, Anna Andreyeva. As she jumped into the trench and began to crouch-run toward the farm buildings, a shell exploded behind her on the field.

Huddled together inside the shelter, Ivanov and his fellow workers braced themselves as more shells pounded the field. When the barrage finally ceased, they emerged into the mist-shrouded night and scanned the darkness for any signs of danger. Finally, they returned to the plot, their resolve undiminished. A cold mist rolled up from below, enveloping the bushes and covering the ground like wisps of cotton wool.

As the first light of dawn bladed through the mist, the group left the scene. The freshly dug black earth stretched for several hundred yards. As the group returned to the farm buildings for food, Ivanov spied a wagtail dive into the field and scour the soil for worms. The men and women gathered in the farm canteen, nursing blisters and warworn spirits. They sat at a long dining table and enjoyed a meager breakfast of flatbreads baked from powdered flour and bran, and a soup made from goosefoot leaves.

III

AFTER A WEEK'S WORK, IVANOV and his colleagues had turned over four acres, working through the nights and sleeping through the daytime. As they became absorbed in their labor, the sense of danger would sometimes fade, until the boom and whistle of an artillery shell restored mortal clarity.

The risks were severe. If he could not sleep, Ivanov would dress and climb the post that overlooked the farm plot. On one occasion he witnessed a scene that would remain crisp in his memory for the rest of his life: in the meadow volunteers were making the first furrow. A pair of thin horses walked slowly along the ground that Ivanov and his coworkers had helped prepare. A woman drove the horses while a man walked behind the plow, pulling on the handles as the earth fell from the plowshare and turned over in the warm air.

Shells fell in the distance. Then one exploded close by on the turned earth. The horses reared their heads and pawed the ground. Ivanov watched the plow lift out of the furrow and the plowshare glint in the sun. The plowman did not flee for cover, however, but pulled the tool backward, pushing the blade back into the furrow while soothing the horses. The animals settled, straightened their backs, and walked on. The shell fragments missed both the plowers and the horses, as if, Ivanov thought, they were protected by a magic spell.

Not every volunteer was so fortunate. One evening, when the botanists awoke for the night's shift, Golikov explained, with a grave face, that the farm had sustained its first losses that day. The spring weather had warmed the soil; it was time to sow. Every opportunity to prepare the land had to be taken; the volunteers would work until the shells began to fall, or until planes began to dogfight above the city, at which point they would lean on the handles of their shovels and roll cigarettes using tobacco wrapped in maple leaves and await the outcome.

That day, walking in a line, each man and woman had been sprinkling handfuls of seeds grasped from the basket slung on their shoulder when a fresh barrage began. The group had run for the trench, but five were caught in an explosion before they made it to safety. Two had died instantly, three were left severely injured. Yevseich, one of the downed volunteers, had managed to fall in such a way to prevent the seeds from spilling from his basket. When Golikov reached his body, he recalled, droplets of blood had soaked into the grain.

Despite these reminders of the risks that faced all citizens, especially those working so close to the front line, the warmth of spring had a galvanizing effect on the city's population. "A thought is forming in fiery letters in my mind," the architect Igor Chaiko wrote in his diary on March 3, 1943. "I can overcome anything. The dangers of the last year and a half have not passed, and war and starvation have not ended. And there are the terrible moments of the things we endured— being so hungry it almost drove you crazy. Yet spring is a symbol of life. The Germans are shelling us again, but the menace is shrinking in the sunlight."

A week later, the plot was ready. Ivanov met with the farm's director to ask if Ivanov and his team had permission to begin planting. The director agreed and suggested that Ivanov take one of the horses to collect the bags of seeds from the Institute. Ivanov untied the horse and began the two-hour walk through the military checkpoints, past the Kirov factory, home to the seed bank.

IV

WHILE IVANOV AND HIS COLLEAGUES planted Vavilov's seeds in the south, the husband-and-wife team of Vadim Lekhnovich and Olga Voskresenskaya began, for the second time during the siege, to remove the potatoes from storage for planting. The previous year's efforts at the nursery by the Krasny Vyborzhets aluminum factory, and the Lesnoye state farm to the north had been a tremendous success. Only one of the rescued potato varieties, known as Tesma—*Solanum tuberosum*—had failed to take. The rest had sprouted through a combination of nutritious soil and careful guarding.

The previous summer Lekhnovich had erected a small cabin and moved onto the plot. He slept during the day and at night walked the perimeter. The nights were light and warm. The young wormwood— *Artemisia absinthium*—had a strong bitter smell, and the bushes cast thick, menacing shadows; from time to time Lekhnovich would strike at the shrubs with his cane—the only protection he carried to ward off thieves—before continuing on his route. By autumn Lekhnovich had stood guard for thirty-eight nights armed with only a stick and a whistle.

"Fortunately," he recorded, "nothing untoward happened."

He had then transported the harvest into the farm's cellar, to again sit out the winter months. Now as moisture turned to steam on the warm soil with the coming of spring, it was time to start planting out the collection's potatoes again.

With the siege ring punctured and the knowledge that the planting plan had worked well the previous year, Lekhnovich and Voskresenskaya turned their attention outward, to help support other growers by sharing their knowledge, refined during the siege months. The previous year they had prepared a pamphlet on how to maximize the yield, propagating tubers from seedlings, cuttings, grafts, sprouts, potato eyes, and even peelings, multiplying the amount of carbohydrates available. The scheme had been a success.

Now Voskresenskaya drew up a plan to train key workers in these accelerated propagation techniques. She scheduled more than a hundred consultations on potato cultivation with individual farmers and authored a fifty-page article for the *Vegetable Grower's Manual*, distributed among all the growers.

Incentivized by the memory of the first siege winter, Leningrad's people had eagerly embraced planting schemes. A schoolgirl named Inna Bityugova worked on the allotments. While cultivating a vegetable garden she became entranced by the mysticism of live botany, as shoots broke the soil and stretched toward the sun. She bought a blank book to document what she saw and on its cover drew the vegetables she was tending—beetroots, turnips, and radishes—anthropomorphizing them with friendly faces and squiggly limbs. Inside she invented adventures for her characters. In one story a vegetable man ran into a building and there found a young, emaciated girl lying on a bed, her arms as thin as sticks, her cheeks gaunt and hollow, her eyes haunted. The character imbued the girl with his strength, as her cheeks filled with color and, finally, she could walk again.

To encourage older gardeners, Voskresenskaya and her husband coauthored newspaper articles to share knowledge with amateur growers, all while supervising continued research on seed development of cancer-retarding varieties. By the end of 1943, the pair had exceeded their targets and provided training to 841 people, delivering forty-five lectures and two radio broadcasts. They had published a monograph describing their discoveries and created a poster on potato cultivation for display around allotments.

Tragedy tempered these successes, however, when one day, during a routine inspection round of one of the allotments where the Plant Institute's collection grew, a German shell exploded in the field. The force knocked Voskresenskaya down. She struck her head in the fall and blacked out. The explosion blew off labels and tore some of the potatoes from the ground, mixing up the ordered rows. When Voskresenskaya came to, she discovered she was almost completely blind.

Her body reduced by severe malnourishment, and now a severe head injury, Voskresenskaya worked more slowly, but no less usefully. When the time came to harvest the specimens that had been mixed up in the explosion that would, eventually, kill her, she listened to the descriptions of each vegetable from her assistants. From their spoken accounts, she identified the varieties so they could be correctly stored according to type.

V

As the summer of 1943 approached, life in Leningrad acquired a fragile semblance of normality. The botanists continued to provide support to those attempting to grow their own supplies of food, who had become more practiced after the previous year's experience. Plants and vegetables swelled in the summer sun as once again residents were given garden plots to tend and weed. The lilacs were particularly fragrant and numerous. In the Field of Mars, women in bright dresses sat reading on warm benches, surrounded by patches of potatoes and turnips. The death rate continued to fall. Residents began to speak of endings and beginnings.

"The Germans now are just a hindrance," wrote the playwright Vsevolod Vishnevsky. "The people have begun to plan the future."

Still, every day was filled with fire and shrapnel. The shelling began in the early mornings, reverberating through the sleeping metropolis, as one diarist wrote, "as if in an empty amphitheater, where the echo comes in layers." The shelling felt capricious and vindictive. "This is the way the Germans avenge themselves for everything—Stalingrad, in the first instance, Kursk, and Karocha, which they lost yesterday."

Yet, as the bombing intensified, so, too, did the people's desire to survive. On April 16, as the writer Vera Inber walked through the streets with a friend, shrapnel began to fall about their heads. In that evening's diary entry, Inber chose to ignore what had become a routine experience of violence and instead focused on the future: "A marvelous day, a spring wind, but not too strong. What a pity it would be to die now, when one wants so much to live."

Leningrad remained, predominantly, a city of women. Female reservists had replaced regular police in directing traffic, a role for which they remained, as one observer put it, "calm, lively, and jolly"—even when one of their number was wounded or even killed by a German shell fragment, to be replaced by the next. Women continued to work in the factories. A quiet city, too. Thousands of Leningrad's apartments sat empty, their tenants dead or exiled. Never in history had Leningrad been so beautiful as in the summer of 1943, noted the novelist Nikolai Chukovsky. To him its emptiness only accentuated its beauty and defiance.

The public had shifted their perspective and priorities. When a Red Army tank unit took cover from German planes among the trees in the Botanical Gardens, one of the curators ran toward the vehicle. He stood in front of the tank, which threatened to tear up the grass and flatten some of the gardens' oldest trees, including a black poplar planted by Peter the Great. The tank commander ordered the curator to move.

"Can't you see there's a war on?" he barked.

"Comrade commander," the curator replied. "These trees have been tended for two hundred years. You will ruin them in a few minutes."

There was a pause, then the officer turned to his crew and ordered them to turn back. "The position is unsuitable," he conceded.

Nature's inexorable work provided daily reassurance. In July, Inber visited the Institute's allotments and was astonished at nature's resilience and indifference to human conflict and suffering.

"Its creations are as beautiful as ever," she wrote. "Such lofty reflections were inspired by things we had taken as a matter of course in peacetime—a calf, strawberries, roses—and which are now simply shattering."

Despite these moments of grace and optimism, the prevailing mood remained strained. Hunger nagged, and while the sounds of shelling had become as commonplace as birdsong, the sights of the resulting destruction continued to provide an arresting reminder of the fragility of life in what had begun to feel like an endless siege. When would war end and, as life returned to normality, the grieving start?

Shelling also continued at the Predportovy farm, where Ivanov and his colleagues planted out around two hundred varieties of seeds. The botanists sowed at dawn, hoisting baskets filled with grain over their shoulders, then sprinkling handfuls of seeds into the plowed earth. As they worked, artillery fire often resounded in the distance in the Pulkovo hills, like spring thunder. The rumble of the guns became routine, as imperceptible to the botanists as the sound of traffic to a city dweller, part of the familiar, unremarkable ambience of daily life.

One day while out planting, Ivanov was knocked down by a blast.

He stood up to see Petrova lying on the ground covering her head with her hands. After another deafening explosion close by, clumps of earth flew, striking Ivanov's face and body like hail from a burst cloud. He smelled the cordite. Then, quiet. Ivanov got up, shook off the dirt and dust, replaced his cap, then walked around the crater. A pool of water had begun to form at the base.

He adjusted the strap of his basket, picked out another handful of grain, and returned to work.

VI

As the sowing season ended, the researchers returned from the various farms in and around Leningrad and regrouped in the cold and silent corridors of the Plant Institute. Outside, the summer sun warmed the pavement, but heat was yet to penetrate the thick walls of the gloomy building, now indelibly associated in the minds of the survivors with so much loss and tragedy.

The optimism felt by Leningraders in the days and weeks after the siege ring had been prized open had been tested by the spiteful artillery attacks, often aimed at civilian targets, and designed to wear down the population. On a Sunday afternoon, a shell hit a tram stop at Sadovaya Street, where locals would congregate to visit allotments. The bodies of the dead lay mixed up with watering cans, shopping bags, and tools. A dismembered arm lay on the pavement, a smoking cigarette still lodged between the fingers, surrounded by splays of carrots and beetroots.

The botanists had kept busy, but it was a disconnected team, working in distinct locations in the area, with their superiors somewhere on the other side of Russia, and the fate and whereabouts of their beloved leader still unknown. If the worst had happened, and Vavilov was dead, what future faced the Plant Institute, even after war? The seed bank was designed to outlast its curators; the point was to establish a continuity beyond the span of any individual. But what would their quiet, hidden acts of incredible sacrifice mean if the Institute survived the siege but not the aftermath?

Ivanov's spirits lifted when he entered the office of his friend and the building's supervisor Klavdiya Panteleyeva and there met Lekhnovich. The three colleagues exchanged weary greetings.

"We've finished the sowing," said Ivanov.

"You must be exhausted," Panteleyeva replied.

"I won't forget that field," he said. "It's not only sweat that people poured into that field. It's blood."

Panteleyeva encouraged her friend to take a day off. Ivanov shrugged off the suggestion. He had too much to do—besides, busyness was a useful way to stave off distress in this purgatorial existence. As the two men left the office together, Lekhnovich mentioned that he had just returned from the Lesnoye state farm, where he and Voskresenskaya were "quietly winning the battle," as he put it, by planting potatoes.

As they walked slowly through St. Isaac's Square—muscle wastage meant that Lekhnovich was unable to board the tram to Lesnoye without support—the sight of women weeding cabbage beds in front of the cathedral interrupted their conversation. They stopped to watch the scene and reminisced about the days when the square was ablaze with flowers, and an ice cream seller could be seen every summer day beside the Astoria Hotel.

A thin child's voice broke the silence.

"Mister, have a drink," the little girl said, offering Ivanov an aluminum cup filled with water from a nearby bucket.

Ivanov drank from the cup, then bent down to talk to the girl, who was helping her mother and the other women with the weeding. He thanked her for the drink and took out a carrot from his briefcase— one of two vegetables Golikov, the head gardener at the farm on the front, had given him earlier that day.

"This is for your kindness," Ivanov said, offering it to her. Then, catching her hesitation, he added, "A gift from a rabbit."

In the summer of 1943, the Nazi botanist Heinz Brücher led an SS commando into Russia on a mission to visit the Plant Institute's various field stations within German-held territory and collect from them Vavilov's valuable seeds. He would then bring these specimens to Lannach Castle in Austria to begin a breeding program of his own.

XV

THE SEED COMMANDO

I

On June 16, 1943, the German botanist Heinz Brücher packed his belongings into a truck parked near the Kaiser Wilhelm Institute for Breeding Research in Munich. Since he had participated in the retreat of German forces from Moscow eighteen months earlier, the twenty-eight-year-old had become a senior ranking member of the SS, Hitler's paramilitary henchmen, whose skull-and-bones insignia indicated a willingness to sacrifice one's life—or to take another—for the Führer.

Brücher and his colleagues Konrad von Rauch, a vegetable expert and SS captain, and Arnold Steinbrecher, an interpreter, were preparing to embark on a journey that would last several weeks and cross hundreds of miles of German-held Soviet terrain. The backs of the two trucks the trio would call home for the duration of the journey sat empty, ready to be filled with treasure.

The day had been long anticipated. Since his return to Germany from the front, Brücher had become increasingly desperate to return to Russia and visit the plant-breeding stations located in German-held territory there. The German army had taken control of around two hundred botanical field stations across western Russia and Ukraine between Minsk and Stalingrad. Many of these institutions had strong links to Vavilov and held thousands of seeds and plants drawn from the world collection in Leningrad. What gems lay dormant in these satellite laboratories?

While Stalin had failed to recognize the value of Vavilov's work, Brücher understood its significance. It was Vavilov's efforts through collecting expeditions and trips that had made wild and primitive crop varieties accessible to researchers. It was his talent and charisma that had, during the 1920s and early '30s, attracted to the Plant Institute the best scientists in Russia—agronomists, botanists, plant breeders, geneticists, cytologists, anatomists, physiologists, agroclimatologists, and geographers—the "kings and queens" of crop research. Brücher wanted to claim the actual fruits of Vavilov's work while there was still opportunity. While the fight for Leningrad ground on, Brücher hoped to capture that which was already within reach.

Other scientists in the Nazi high command shared Brücher's envious perspective. Two years earlier the director of the German Institute of Biology, Fritz von Wettstein, had first raised the idea of creating a new plant-breeding center based on the Soviet-held plant matter claimed by the German advance. That year the Nazi command issued the order to appropriate all the Institute's experimental stations within German-held territory. But the seizure of the Russian stations had been chaotic. Competing government departments jostled for control, complicating the process. By 1943, there had not yet been a coordinated effort to capture the plant material and bring it to Germany for study.

On February 2, 1943, seven days after the doctors at Saratov pronounced Vavilov dead—a fact unknown to almost anyone in Russia, and nobody overseas—German troops had surrendered outside the city of Stalingrad. When Brücher learned of this critical defeat, he felt a renewed sense of urgency. The fall of Leningrad that would grant him access to the world collection of seeds seemed increasingly improbable. Moreover, if Soviet troops retook Crimea and Ukraine, Germany would lose access to a clutch of experimental breeding stations along with the vital plant material derived from the primary collection.

Brücher's anxiety was fueled by loyalty to the Nazi cause. "Having recourse to these primitive strains of crop plants collected by Vavilov, in the current stage of plant genetics, is all the more significant," he wrote in one report. By "crossing the genes that were selected under the severe selection conditions of nature itself," German scientists could use Vavilov's collection to improve their crops' resistance to "cold, drought, and pests." Like Vavilov, Brücher believed that these seeds and plants held in their genetic material the secret to food security for his nation.

He was also motivated by professional interest and frustration at what he saw as the incompetence of his superiors. Brücher lamented that months of valuable time had been squandered. "Unfortunately, we Germans made major errors," he noted. "During the forward push and the battles in the east, I had observed that the material of numerous Russian breeding institutes and the material brought together of irreplaceable value were destroyed." In Brücher's view, these acts had been conducted either "in ignorance" or as thuggish acts of vandalism against enemy assets.

The Soviet victory at Stalingrad had marked a turning point in the Eastern Front, halting the German advance and demoralizing their army. It shattered the myth of German invincibility and boosted Soviet morale. The tide of the war had begun to shift. As the German forces began to retreat toward western Russia, these botanical treasures might be lost forever, not through negligence, but through military fiasco.

Brücher had some clout; he served on Himmler's personal staff in the SS Ahnenerbe, a research unit composed of scholars and scientists from various academic disciplines devoted to proving the theory that the German people descended from a racially superior ancestry. The Ahnenerbe conducted archaeological excavations, anthropological studies, and ethnographic research, and the group's findings were often disseminated in Nazi propaganda and used to justify the regime's policies. Brücher's position within this renowned group granted him access to the higher echelons of Nazi leadership, and with it a certain capacity to get things done.

On May 8, 1943, Brücher met with SS general Oswald Pohl, head of the finance and administration bureau. Brücher presented Pohl with his plan for a botanical raid mission. The pitch was straightforward and practical: he would personally tour eighteen test stations across Ukraine and southern Russia as part of an SS *Sammelkommando*— a collecting commando unit—on Himmler's behalf. Brücher and his team would rendezvous with German forces in each area, who would provide additional support and, if the Soviet botanists attempted to prevent the removal of seeds and plants, physical backup.

While the German army continued its plodding retreat across Russia's grand expanse, Brücher and his men would press forward in the opposite direction, making a determined dash to claim the genetic plant material before it was too late. Once the seeds had been procured, the unit would bring the looted plant matter to a castle in the remote Austrian town of Lannach. There the spoils could be combined with another collection from an Ahnenerbe expedition to Tibet five years earlier and together form the basis of a new SS Institute of Plant Genetics.

Himmler told Brücher he was to collaborate with the Kaiser Wilhelm Institute for Plant Breeding in Müncheberg, and the newly formed Institute of Cultivated and Wild Plants, situated in the estate of Klosterneuburg Abbey, on the northern outskirts of Vienna. Together, the institutes would increase their efforts to seize everything they could from the Soviet plant-breeding sites, using Vavilov's collection to form the basis of their own rival seed bank.

In mid-June, Brücher and his two colleagues began the long drive eastward, carried on the winds of Nazi support, on the first recorded journey of state-organized biopiracy. On the long drive from Munich to Ukraine, a cloud of concerns blotted out the summer sun from Brücher's mind: In what state would he find those experimental stations that had survived the German invasion, but had endured two years of occupation and neglect? With what resistance would the commando unit be met?

II

THE FALL OF STALINGRAD HAD inspired a campaign of spiteful artillery attacks on Leningrad from German positions. Throughout June

THE SEED COMMANDO   237

hails of artillery shells provided a daily reminder that the city's people were not yet free of the specter of occupation. Some residents chose to remain indoors in the cool and comforting shadows of their apartments. Those who ventured into the sun often found their routines broken by another barrage. The randomness and unpredictability of German artillery encouraged superstitions—a preferred side of the street, taken at a certain time—modest ways to relieve the singular feeling of helplessness induced by a hail of explosives.

One day Nikolai Ivanov permitted himself a break from his work and took his wife to the cinema to watch *The Swinemaiden and the Shepherd*, Ivan Pyriev's black-and-white musical comedy. As the couple returned home after the showing, the sirens blared their warnings and they ducked into the nearest shelter. Inside, men and women loitered as if waiting for buses, unpanicked by what many Leningraders now considered to be a vindictive campaign of terror by an enemy on the verge of retreat.

When the all clear sounded, the couple emerged and walked along Nevsky Prospekt. Despite the extensive damage to buildings, and the boarded shop windows, the street was clean. To celebrate the act of survival, Ivanov entered No.1 Grocery Shop and purchased some non-ration food: a can of Olivier salad.

Ivanov cherished these trips as a chance to reconnect with his wife and experience a different sort of life inside Leningrad. Since April he had spent most of his working days close to the front, plowing and planting on the Predportovy farm. At night he had joined the guard duty, ensuring locals did not steal into the field and make off with vegetables. That summer, the seed bank's director, Johan Eichfeld, had returned by plane from Krasnoufimsk to Leningrad, flying across Lake Ladoga, and, once inside the city, visited the farm to inspect Ivanov and his colleague's efforts. Now, as Brücher and his commando unit drove toward Ukraine to steal the seeds preserved in the experimental stations, the time had come to harvest and renew some of the primary collection.

When Ivanov returned to the farm, he and his colleagues learned there had been losses while they were away. A thief had slipped past the guard one night and made off with some onions; some of the cabbage heads had been taken, too. And while a few of the samples the team had planted from subtropical countries and high-altitude regions had failed to germinate, the planting had been an overall success.

More painful was the news of further deaths of farmworkers by German artillery fire from nearby enemy positions. Six had died while harvesting crops under cloches. Perhaps the German watchers had noticed the bustle of activity in the field or caught a glint of sunlight from a glass pane as a cloche was lifted. As the shells had begun to fall on the site, Ivanov's friend Ivan Golikov, the farm's chief agriculturist, gave the order to abandon the site and take cover. His words were cut short by a shell. An eyewitness described how Golikov's lifeless body lay on the ground, his arms outstretched, as if protecting the shoots. By the time Ivanov arrived, the six were already buried at the nearby Krasnenkoye Cemetery.

Ivanov and his colleagues harvested the Institute's allotment without Golikov. They scythed every furrow, threshed the grain by hand, winnowed it in the breeze, then returned the seeds to the Institute, to be sorted and stored in the autumn. The scheme had been a costly success.

III

THE TWO SS TRUCKS TORE determinedly into Ukrainian territory. Brücher, young and fiercely ambitious, was eager not to waste time. Each day the Red Army made gains on German-held territory, increasing the risk that valuable plant matter might be lost. And each experimental station contained different seeds and different plants, the results of experimentation to breed plants suited to the varying conditions. This archipelago of stations had worked for decades to produce seeds optimized for local conditions, and each location provided a different section in Russia's enviable genetic reference library.

When Vavilov was not trekking across mountains or through valleys, he had spent much of his time attending to and overseeing these experimental field stations. He had established the first on the out-skirts of Leningrad, in the town of Pushkin, shortly after he assumed leadership of the Institute in 1921. There, on the grounds of the tsar's summer palace of Tsarkoye Selo, stood a replica of an English country house—a gift from Queen Victoria to her godson, the ex–grand duke Boris Vladimirovich. The building materials had been sent, beam by beam, tile by tile, from England and used to construct a home unlike any other in the region. After assuming control of the building from the Communists, who occupied it during the Revolution, Vavilov's team grew seedlings in the expansive greenhouse on its grounds. Be-fore he was killed at the front in the second month of the invasion, one of Vavilov's team, Fyodor Prokofyev, built a mill in the building for processing flour and baking bread.

The stations at Pushkin and Pavlovsk enabled Plant Institute staff to easily transport seeds and samples for experimentation from the center of Leningrad to the suburbs. Vavilov also oversaw stations far-ther afield, each one strategically located in areas that enjoyed differ-ent soil types and climates, to help the botanists understand which plants and hybrids might flourish across Russia's vast and diverse land-scapes. In 1923 he oversaw the establishment of the Polar station, near the town of Kirovsk within the arctic circle—a natural laboratory for studying crop variability in freezing temperatures.

In 1929 he oversaw the establishment of the Far East Experimen-tal Station, twelve miles from Vladivostok, dedicated to the study of fruits and berries, aboriginal forms of plum, apricot, magnolia vine, honeysuckle, and Amur grape. The Volgograd Experimental Station, in the floodplains of the Volga, enabled botanists to study vegetable- and potato-seed production under intensive irrigation conditions. The Krymsk Experimental Station, founded in 1935, studied green pea, maize, tomato, pear, eggplant, cucumber, apple, plum, peach, cherry, strawberry, and apricot.

At the Kuban Experimental Station, situated in the steppes of Krasnodar, Vavilov's botanists studied maize, sunflower, and castor oil plants and conducted immunological research on wheat, barley, chickpeas, and sunflower, and flax resistance to fusarium wilt. At the Maikop Experimental Station they bred onions, peppers, eggplants, cucumbers, cabbages, garlic, carrots, and red beetroot.

While Brücher dreamed of striding through the doors of the Plant Institute in Leningrad, or those of the glamorous Pushkin Station, and rifling through the tins and crates that contained seeds gathered from the farthest reaches of the world, he would, for now, settle for the chance to take what was available to him. He planned to gather all the viable plant matter collected by Vavilov and his teams during the past two decades, and take them to Austria for evaluation, to ascertain the best way they could be used for the betterment and survival of the German people.

After Vavilov's arrest and the German invasion, work continued at the field stations—the degree of their effectiveness and industry often set by the leader. In the spring of 1942, for example, the scientists at the Polar station distributed 71 tons of cancer-retardant potatoes, almost 224 pounds of best-quality seeds from fruit cultivars, a quarter of a metric tonne of best-quality forage seeds, and more than four metric tonnes of grain cultivars to local farmers. Other stations, especially those within German-held territory, cut off from Soviet funding and leadership, had fallen into disrepair.

The commandos' first stop was Oleksandriya, near the city of Bila Tserkva in central Ukraine. While the laboratories and scientific facilities had not suffered any war damage, the state of scientific work at the station confirmed Brücher's fears. The German occupiers had installed a local farmer as director of the station. The farmer specialized in animal breeding and had no scientific background in botany or plant breeding. "Consequently," Brücher wrote in his report, "there was no more scientific research carried out at the station." The several hundred acres of land attached to the station had not been used for scientific study, but merely to grow cereals and vegetables.

Next, Brücher's commando unit drove fifty miles east to the city of Myronivka, south of Kyiv, one of the largest breeding stations anywhere in the Soviet Union. When he arrived, Brücher introduced himself not as a military officer but as a scientist. He hoped, perhaps, that by approaching each station as a professional peer, rather than a soldier, he might smooth his assuming control of the plant matter. There was never an attempt to gain entry to an experimental station by force. Even so, Brücher carried with him the full backing of his superiors. If required, he could produce letters of authority to requisition seeds at any station signed by Hans-Adolf Prützmann, one of the highest-ranking SS leaders in the occupied Soviet Union; Adolf von Bomhard, the Nazi-installed head of police in Ukraine; and Ludolf von Alvensleben, police chief of the Crimea.

Brücher was pleased to find the experimental station in Myronivka in a clean and orderly state. A German special leader had assumed command of the institution and had convinced the Russian scientists to continue their work under his direction or, at least, inspection. The fruits of Vavilov's research work were obvious: the world collection had allowed the Russian scientists to breed healthy, hardy plants, the kind that promised to increase yields and, it followed, national food security. So successful, in fact, it had attracted the attention of other interested parties.

As Brücher produced his paperwork and explained his mission, he was dismayed to learn that he was not the only German planning to appropriate the seeds. Representatives from a private German plant-breeding firm had already visited the station with a request to take over the premises, presumably to begin growing seeds they could sell. Brücher had feared that administrative dillydallying might have spoiled the Nazis' chances of securing the best and most viable plant matter, but he had not considered that he might be competing with German businesses.

From there, Brücher's unit began a tour of Ukrainian stations. His hunch that the plant material and staff held in the westernmost field stations of the Plant Institute had not been evacuated to the Urals proved correct. The agricultural stations were mostly intact, albeit often neglected. There was no resistance from the Russians working at these laboratories, some of whom may have favored their chances with the Germans following years of disregard, underfunding, and persecution under Stalin's regime.

The scientific ethics of plant science meant that Vavilov and his colleagues regarded their seed collections and plant materials as property of the international scientific community and, beyond that, humankind. The scientists and academics Brücher met were often more concerned with the preservation of their plants than the allegiance of those looking after them. For this reason, many chose to risk accusations of collaborating with the enemy.

At the Agricultural Institute in Uman, south of Myronivka, Brücher's team were warmly met by the director, Professor Chrennikov, who had studied in Germany years earlier. Chrennikov complained that the Soviets had "no understanding of the importance of his breeding station," which had opened in 1814. Even before the outbreak of war, he explained, he had feared closure.

By contrast, at the Research Center at Kherson, in southern Ukraine, Brücher found the station in rude health, under the directorship of Professor Cherkasov, who had been willing to cooperate with the German Military Administrative Council. More than five hundred different strains of grain, including rice, were being studied and bred in the station—providing a sizable haul for Brücher's crew.

The Nikita Botanical Garden, close to Yalta and one of the oldest botanical gardens in Europe, presented a more disheartening scene. "This once-famous Russian garden made a desolate impression from a scientific point of view," Brücher reported. Before the war, there had been more than a thousand vine varieties, more than five hundred peach varieties, three hundred varieties of fig, and numerous cultivated and wild plums, as well as unique crosses between peaches, almonds, quinces, and pears. In Brücher's view, the occupying Germans had squandered this scientific bounty. "The lack of robust German direction was particularly obvious," he wrote.

As he interpreted for Brücher, Steinbrecher came to better understand the character of his leader. Brücher was clearly meticulous and intelligent, but "very ambitious, demanding, forceful, and autocratic."

The young commander was polite until someone stood in the way of his plans, at which point he could turn belligerent, even cruel. As the seed-collecting commando moved from station to station, the young Nazi's irritation toward German mismanagement curdled into anger.

Repeatedly he found his efforts met with opposition, not from the Soviet scientists, but from the quarrelsome officers who had assumed control of these regions. At the Synelnykove station, close to the Black Sea, the German in charge of the experimental station flatly refused to hand over any seed samples to the commando unit "despite," as Brücher wrote, being shown "the fully authorized permission from *Obergruppenführer* Pohl and *Obergruppenführer* Prützmann."

The refusal was especially galling to Brücher as, after interviewing the scientists who worked there, he had established that the facility contained thousands of strains of barley, wheat, oats, maize, and rubber plants, all of which had been grown directly from Vavilov's world collection. Unmoved by Brücher's pleas, the director explained he had received written instruction "that no seeds were to be handed over to the Collection Command currently in Russia."

While Brücher seethed, the Germans who had arrived before him deployed delaying tactics and even, according to Brücher's report, tried to hide material. At the Aleksandrovsk station, Brücher was incensed to find that occupying staff had wanted to demonstrate the superiority of German wheat and planted sixty acres with the Heines variety—a pointless waste of agricultural space from Brücher's perspective. "The result was crushing," he recorded. The German summer wheat yielded less than half of what had been expected. "[It] remained a long way behind the yield of the Russian country varieties."

At the Batai-Berg test station in Kyiv, Brücher's team discovered a vast array of valuable seed samples, many planted out across two hundred and fifty acres of land. The subcollection comprised more than ten thousand European varieties, and more than forty thousand from Asia and America. There were too many samples to load into the trucks, so Brücher made an agreement with the Soviet director, Professor Vyacheslav Savitsky, that samples of every kind would be collected and taken in late autumn to the castle in Austria. But the German authorities delayed deliveries until it was too late. When the Red Army recaptured Kyiv in November, the seeds still had not left the experimental station. "The material could no longer be brought to safety," mourned Brücher.

The German guards at three of the eighteen field stations Brücher and his team visited refused to part with seeds, despite his insistence that he had the rank and authority to do as he pleased. With each refusal, he issued a menacing rebuttal, saying, "You bear full responsibility for the consequences of this refusal and may be brought to justice before the SS authorities."

Brücher took pleasure in naming the individuals in his official report and follow-up letters: Special Leader Dr. Denkhaus, leader of the plant breeding ring of Dnepropetrovak; Dr. Flachs, the officer responsible for seed management for the Ukraine *Kriegsverwaltungsrat*; and Special Leader Grey, whose refusal on July 31, 1943, to hand over plant material to the commando, resulted, Brücher wrote, in "all of the material [being] lost to German breeding."

Dr. Denkhaus so enraged Brücher that in his report he impugned his compatriot's intelligence, writing, "The seven Russian academics present [at Synelnykove] were intellectually far superior to the German director of the plant breeding ring, Denkhaus."

Despite the setbacks, which Brücher viewed as the result of incompetencies of German middle management, he assured his superiors that the mission had been a triumph. "The SS Collection Command's seed material secured . . . has a very special significance," he recorded. After a three-month-long tour of Soviet breeding stations, Brücher and his colleagues turned back with trucks loaded with seeds and samples, a stolen trove that could now be used as the basis for his own "world collection" of seeds in Germany.

He was aware, too, of the haul's vulnerability. An outbreak of disease could wipe out the spoils of the trip. For this reason, he urged his superiors, the German scientists needed to begin their work growing the seeds, testing hybrids, and increasing resilience. By combining

Vavilov's seeds with those from the Ahnenerbe expedition to Tibet, the German scientists had the opportunity to continue and even improve upon Vavilov's pioneering work.

With around forty thousand samples collected, Brücher and his men began the long drive to Austria, to Lannach Castle, a sixteenth-century stronghold that would now become the SS Institute for Plant Genetics, a fortified rival to Vavilov's seed bank in Leningrad.

There, as Brücher concluded in his report, he and his staff would begin the "breeding of cold-resistant, drought-resistant, and fast-growing crop plants for the eastern territories." Vavilov was dead, and his collection poised to feed Russia's enemy.

At 8:00 p.m. on January 27, 1944, the surviving botanists from the Plant Institute
gathered with thousands of others in parks and on bridges and embankments
for the official victory salute: the siege was finished. A few months later Vavilov's
brother, Sergei, visited the seed bank. That night in his diary he wrote, "If only God
and souls could be brought back inside these walls."

XVI
TWENTY-FOUR SALVOS

I

Sᴇʀɢᴇɪ Vᴀᴠɪʟᴏᴠ ꜰɪʀꜱᴛ ʜᴇᴀʀᴅ ᴛʜᴇ rumor that his older brother was dead in the summer of 1943. Worrying about Vavilov's well-being and whereabouts had become a habit for Sergei, not of consolation but of self-recrimination. Three years earlier, Maria Shebalina, one of Vavilov's supporters at the Plant Institute, had visited Sergei at his office at the Russian Academy of Sciences on Vasilyevsky Island. In the document she passed him, she outlined in detail all the material benefits the Soviet Union had derived from the director's expeditions, his experimental work and research. Shebalina explained that, if the document could reach the right person in authority, it might help inspire restraint, even a pardon for Sergei's brother.

Scared that he might be caught up in Vavilov's predicament, Sergei had hardened his heart and dismissed Shebalina. There was, he thought, nothing more to be done—at least, not without risking his and his family's own lives and freedom. "We can hardly hope to do anything about it," Sergei told Shebalina. It was the first of what Sergei had come to view as a series of appalling disloyalties toward his older brother.

Two years later, when the British had inquired as to Vavilov's whereabouts, Sergei had colluded with the Soviet state in maintaining the fiction of his brother's freedom and well-being. He faked his brother's signature on the document designed to mislead Vavilov's international friends and supporters by pretending that all was well. These were acts of survival as much as of betrayal, actions undertaken to avoid being labeled a spy, a wrecker, or an enemy of the people. Still, the private strain of these treacheries had grown. Life without his brother, Sergei had written in his diary, had become "unbearably hard." At night he was haunted by visions of Vavilov, who would appear in Sergei's dreams "skinny" and covered in scars of "baked blood."

Sergei suspected that, even if his brother survived the war, he would not survive the inexorable squeeze of the Soviet gulag. In the spring of 1942, Sergei had written in his diary, "Maybe he is dead." Sergei could not visit Saratov to discover the truth himself—he could not leave his post at the Optical Institute, which had been evacuated to Yoshkar-Ola, four hundred miles to the north of the prison. So, he arranged for his nephew Oleg—Vavilov's eldest son—to travel in his stead.

Oleg was twenty-five and had followed his father into the sciences as an employee of the Lebedev Physical Institute of the Academy of Sciences. Sergei arranged for the relevant travel permits for his nephew, who took the train from Moscow to Saratov, the reverse of the first leg of the journey his father had taken in 1921, when he left the city with his crates of seeds to establish the seed bank.

When he arrived in Saratov, Oleg first visited his stepmother, Yelena, and his half brother, Yuri. Both were still unaware that they had been living in the same city as Vavilov for some time now. Not wanting to arouse suspicion Oleg arrived unannounced. The house was cold and sad. Barred from employment through association, mother and son had survived on a daily bread ration of eleven ounces of black bread each, and whatever money they received in the mail from the guilt-ridden Sergei. On one occasion, their house had been broken into after a neighbor watched Yuri drag a sack of potatoes into the basement of the house, purchased with money sent by his uncle.

Oleg explained that he had come to Saratov to inquire as to Vavilov's whereabouts. Yuri, who was nine years Oleg's junior, agreed to accompany his half brother to the NKVD headquarters at Gray House the following morning. The next day the pair arrived at the forbidding reception hall on the first floor of the building. Fearing the worst, Oleg insisted that Yuri wait on the far side of the room while Oleg approached and knocked on the shuttered window.

The shutter rattled up and Yuri watched Oleg lean in. The conversation was brief. After a few moments the shutter clacked down, and Oleg returned to his half brother. He explained that the official said he could not help and had told Oleg to inquire at the NKVD headquarters in Moscow instead. The pair returned to Yuri's home. Oleg bade his goodbyes, promising that he would return to the capital to discover more. When Yuri offered to accompany his brother to the station, Oleg insisted he must go alone. He would be in contact, he assured the teenager, just as soon as he had news to share.

The young men parted ways. Oleg, however, did not go to the train station. He peeled off and doubled back to the post office. There he composed two telegrams, one addressed to his young wife, Lidia, and the other to his uncle Sergei.

Both messages read, "Died, January 26, 1943."

II

LANNACH CASTLE, IT SEEMED TO William Denton Venables, had been misleadingly named. No moat or keep, no looming gatehouse or jagged battlements to hold back European invaders. This sixteenth-century Austrian castle situated on a small hill to the east of Lannach town center looked more like a mansion than a fairy-tale fortification: a two-story yellow stone house with a modest courtyard and scattering of cobwebbed outbuildings. The floorboards had warped from where the roof had been leaking for several years. The rooms and corridors were damp, the paint was bubbled and curled. In places the plaster had begun to break away.

Still, for Venables, a trained botanist who in 1940 had joined the Fourth Queen's Own Hussars, the distinguished cavalry regiment in which Britain's prime minister Winston Churchill once served, it was not the building that excited him so much as the prospect of a break from Stalag XVIII-A. He and a small group of other prisoners had been seconded from the overcrowded, disease-riddled camp, forty miles to the west of Lannach—although none was quite clear yet on what basis they had been picked, or for what purpose.

For Venables, who had grown up in the small town of Whitchurch in Shropshire, near the border between England and Wales, it had been an arduous war. Two years earlier on April 28, 1941, his regiment had met the Nazi invaders in Kalamata in Greece, during Operation Marita. The Germans had overwhelmed the Allied resistance. During the chaotic evacuation of the British divisions, German paratroopers tried to secure the bridge over the Corinth Canal. Venables and his compatriots repelled the paratroopers long enough for British engineers to blow up the bridge, but it was a temporary holdback. Overwhelmed by opposing numbers, the regiment surrendered, and Venables and more than four hundred other British soldiers were taken prisoner and escorted to a nearby transit camp.

Venables wasted no time mounting an escape. That evening, while leading a group of prisoners sent to collect water for the camp under armed escort, he darted into a nearby vineyard, where he reasoned he could lose his pursuers among the vines. A Nazi guard chased him for a short while, but did not fire his weapon. Having lost his pursuer, Venables rendezvoused with three officers and seven soldiers from his unit. The men made a plan to borrow a boat and sail to a neighboring island. They found no willing coconspirators, however. The risks for the local population of aiding escapees were too great; the most any beleaguered local could offer was some food. Even then, supplies were limited.

After a month spent living off the land and the kindness of strangers, Venables and his compatriots were, as he later recorded, "practically starving." On May 28, 1941, exactly a month after he was captured and escaped, the young trooper surrendered to a German patrol: better to capitulate and live to escape another day than to perish in the undergrowth. He was taken to a transit camp in Corinth, a hundred miles to the north, and from there to Maribor in what is now Slovenia. Finally, in March 1943, Venables arrived in Austria, the newest captive at the typhoid-racked POW camp known as Stalag XVIII-A.

When his captors learned Venables was a botanist, he was assigned to work at a local farm. In June 1943, three months after he arrived and while Heinz Brücher was still leading his seed-collecting commando unit on its looting mission across Ukraine, Venables staged a second escape attempt. This time he slipped away from the field where he was working unseen and walked to Graz railway station. Venables had enough money to purchase a ticket to Klagenfurt, on the eastern shore of Lake Wörthersee.

The botanist spent three days hiding in the countryside, then took the train to Innsbruck. En route, he was asked to show his civilian papers. Unable to produce any documentation, guards escorted Venables to an Innsbruck military guard room where he was questioned, then, finally, returned to the camp from which he had escaped. There he remained until, unexpectedly, the guards informed him he was being sent to Lannach.

There the POW met the prickly and determined Heinz Brücher. Despite the castle's mounting dilapidation, the grounds were well suited to the German botanist's plans to build on the work of Vavilov and his team. The castle grounds comprised 90 acres of arable land, 44 acres of meadowland, and 150 acres of forest. With its hot summers and long autumns, the local Styrian climate was ideal for experimental growing. The village of Lannach sits at the foot of the Coral Mountains. Brücher envisioned a network of cultivation plots located at different elevations on the mountainside to maximize results.

The German explained to Venables that he planned to select seedlings from the Soviet and Tibetan stocks and crossbreed them with European strains to develop hybrids that would thrive at higher elevations and in colder climates. The seeds and plant samples the commando had collected from the Soviet Union were added to the wheat, barley, and oat samples that had been collected during an SS expedition to Tibet in

1938, led by Himmler's protégé the zoologist Ernst Schäfer. Since some of the seeds in the collections had been harvested from high altitudes, where the growing season was short, Brücher also hoped to be able to produce fast-ripening varieties. Drawing inspiration from Vavilov's example, Brücher eventually planned to produce strains of grain that could grow in the arctic.

Brücher was pleased to have an engaged professional botanist at his service in Venables, particularly as the Englishman spoke fluent German. The pair developed a rapport. Both men were twenty-eight years old—only three months separated them—and their shared interest in botany provided a focus away from differences of nationality and ideology. Soon they began to collaborate, not as captor and captive, but as colleagues. After the day's work Venables was not required to return to the POW camp, but could retire to the castle's former granary rooms, which now served as dormitories.

On November 1, 1943, Himmler appointed Brücher head of the Institute for Plant Genetics, SS-Versuchsgut Lannach. It was to be a five-month posting, on loan from the Wehrmacht, that he would hold until the end of March 1944. Brücher, eager to demonstrate the benefits of his commando's intrepid seed-salvaging project, worked quickly. While he oversaw the refurbishment of the castle and its outbuildings, the botanist planted out 202 lines of wheat, 18 of rye, 74 kinds of barley (including 12 varieties of *Hordeum bulbosum*), and 28 kinds of oats. His collaborators included Volkmar Vareschi, a thirty-seven-year-old Austrian botanist and plant ecologist and expert in Venezuelan lichens and ferns, and twenty-three civilian employees.

Of his workers, Venables was Brücher's favorite. For Brücher, here was an expert with none of the political baggage or rivalry of his contemporaries in German science. For Venables, here was an escape from the discomforts of his camp, the chance to work with plants and soil, an escape from the psychological demands of war, and, perhaps beneath that, the hope that, should the Allies lose the war, he had a powerful supporter in the SS. Whatever the motivation and authenticity of each man's feelings, the alliance worked. Brücher would later describe the British trooper as "my British wartime collaborator." In a spirit of shared botanical enthusiasm, they set about the task of planting out Vavilov's stolen seeds.

III

SERGEI VAVILOV WAS DISTRAUGHT WHEN he received his nephew Oleg's telegram. The news of his brother's death felt inevitable, but until this confirmation there had been room for hope. No more. Sergei told his wife, Olga, the news. She let out an "unforgettable" scream. "My own soul is frozen and become like a stone," he wrote in his diary. "I long for a quiet, fast, invisible death." When a friend visited the couple, Olga wept while Sergei retired to his room, too depressed to entertain a guest. Eventually, when he emerged, his friend was shocked to see Sergei's face "close to black," darkened by blood vessels near the surface of the skin. "He sat like a person who is somewhere else," the friend recorded.

While grieving, Sergei remained responsible for organizing the transfer of the Optical Institute to Moscow, where he and other physicists had been called to focus on nuclear physics and how the Soviet Union might weaponize atomic power. Despite the urgency of the work, Sergei functioned, he wrote, "like a robot." Work became a stultifying routine. Whenever he caught his reflection in a mirror or shop window, Sergei caught his brother's likeness and winced. He began to wear his brother's overcoat.

Two months later, in October 1943, agents at the NKVD Moscow headquarters summoned Sergei to sign a paper confirming his brother's date of death. Any private belief the physicist had indulged that the authorities in Saratov had been mistaken dissipated.

"The last thin thread of hope is torn," he wrote.

After he signed the paper, Sergei began to send Yelena and Yuri money to help them survive the coming winter. To spare them the anguish of what they surely suspected, Sergei did not pass on details of Vavilov's death, or even the fact that he was buried close to where they lived. Perhaps Sergei feared it would cause a distress from which Yelena would not recover.

Unaware of his father's fate, Vavilov's younger son, Yuri, continued to provide for his mother, who had come to suspect the worst. He had, it seemed, his father's talent for working with the land. The following spring, he planted cucumber seeds, which he watered each day and fertilized with horse manure he collected from the nearby road. In time the plants produced long and succulent vegetables—too many for them to eat themselves. He took the surplus to the local market in Saratov and sold them to raise funds to support his heartbroken mother.

IV

Abram Yakovlevich Kameraz, the potato specialist, who had rescued his precious South American varieties under German artillery fire, was war-weary. Since leaving his pregnant wife in Leningrad to join the 123rd Rifle Division of the Soviet Union, he had defended the city from various positions, first with the 23rd Army, defending the northwestern approaches to Leningrad, then with the 67th Army, as part of Operation Iskra. During this time, Kameraz had fought as a rifleman, a mortarman, and a gatherer of military intelligence. More recently, as a fluent German speaker, Kameraz had been employed as an interpreter, helping with the interrogation of captured prisoners. He had been wounded twice, once seriously. A restless patient, while recuperating he had authored the combat history of his regiment.

As the end of 1943 approached and Hitler's generals attempted to persuade their Führer to withdraw from Leningrad, Kameraz caught a picture of a demoralized German force from his interviews with POWs. A deserter from the German garrison at Novgorod, who had decided to present himself to the enemy before he was killed, explained that his officers now spent all their time drinking and gambling, while the infantrymen put their faith in "some destructive weapon that has so far been kept a great secret."

"Russia is too big for us to defeat her," mourned another.

In interview after interview Kameraz and his colleagues felt the rumbles of change. Perhaps, soon, he could return to Leningrad and discover who among his colleagues had also survived.

On October 6, 1943, General Govorov began to prepare plans to break the siege, a three-pronged attack from the Oranienbaum bridgehead in the west, the Pulkovo Heights in the south, and Volkhov in the east. The offensive, known as Operation Neva, would launch in the winter weeks, when the ice enabled his troops to move more easily.

In the weeks that followed, guns, Katyusha rocket launchers, tanks, and aircraft massed inside the city. More than a thousand freight cars laden with shells arrived, the munitions they carried adding to the piles produced by Leningrad's own factories. Onto these shells the workers had stenciled inscriptions for their eventual recipients: FOR THE BLOOD OF OUR WORKERS; FOR OUR CHILDREN'S ANGUISH; FOR OUR MURDERED FRIENDS.

By the time preparations were complete, Govorov's forces outnumbered the Germans threefold in infantry, fourfold in artillery, and sixfold in tanks. It was the largest concentration of firepower yet seen in the Soviet Union. Then, on November 6 the Red Army liberated Kyiv, in Ukraine. The victory only increased the sense of coiled power felt by the liberators-in-waiting.

Revolution Day. New Year's. As Abram Kameraz and his compatriots in the 123rd Rifle Division joined the Front Subordinate, one of six gathered armies poised to liberate Leningrad, it became clear to the botanist that the decisive moment in the battle for his city had arrived.

V

AT DAWN ON JANUARY 14, 1944, a Russian shell screamed through the fog to land on the frosted German positions. So began an infernal bombardment from Oranienbaum. At 10:00 a.m., the massed infantry of the 48th, 90th, and 131st Rifle Divisions advanced on the German defensive positions. General Govorov observed the assault from a command post on Kolokol'nis Hill.

"We're living through hell," a German officer wrote to his wife that evening. "I can't describe it . . . wish me luck."

The next morning, at 9:20, the botanists in the seed bank were at their desks when the windows in their offices began to shudder. They put down their books and papers and walked outside to join the crowds staring into the sky. The city echoed with booming.

"It's begun," cried one onlooker. "It's begun."

In an hour and forty minutes, more than two hundred thousand shells fell on German positions at Pulkovo.

Inside the Plant Institute, plaster crumbled from the walls while the light fixtures swayed with each tremor. At the nearby Sudo-mekh Shipyard, which specialized in producing submarines, a work-shop crashed to the ground. In her diary one onlooker, Vera Inber, recorded, "A continuous rumble, of greater strength than we have heard before, shakes the air. . . . What is going to happen? Can it be that we shan't succeed this time either? But everyone believes that all will be well."

During the next few days, a procession of Red Cross buses col-lected the wounded from Leningrad's railway stations to take them to the hospitals.

"So long as their blood is not flowing in vain," wrote Inber.

VI

ON JANUARY 21, 1944, SEVEN days after the assault began, the Red Army reclaimed Mga, the railway station whose capture had pre-vented the trainload of seeds from the Plant Institute from making it out of the city. That morning Inber received a telephone call from the Writers' Union telling her to be ready: in an hour she would be collected for a tour of the newly liberated spaces. She was to record what she saw.

The journey commenced at the Kirov works, near the farm where Ivanov and his comrades had tenaciously planted the seeds the previous spring amid the onslaught of German artillery fire. Inber's words captured the desecration: "Everything is plowed over by war. Barbed wire, bunches of electrical wiring, ditches, the brown debris of destroyed houses everywhere. Long tongues of soot radiate out on the snow from bomb craters, the strength of the flames can be determined by them."

Only the walls of the clinic at Forel Hospital remained. The typewriter factory, which the Germans had used as a fort from which they fired their shells, now stood as a ruin. Inber noted the invaders had added German inscriptions, arrows, and even cartoonish illustrations to the road signage. Even the most inconspicuous bridges had not been spared. The driver proceeded with caution; a pickup truck in front of Inber's vehicle had struck a mine, leaving the driver sitting by the roadside, tributaries of blood running down his face.

"Where the hell do you think you're going?" the injured man shouted at the driver of Inber's vehicle. "Can't you see for yourselves what is happening?"

On the sides of the road Inber noticed round mines arranged on planks, defused by Red Army sappers to make way for friendly vehicles. "Other mines, longer with feathery tails, are piled in heaps and look like dead fish."

As the truck crept onward, Inber peered through the window at the battered parks. Nature bore the scars of human conflict. Trees stood cracked and splintered, their branches cleaved off or hanging at excruciating angles.

"One tree," she wrote, "battered, frightening, its bare branches uprights, seemed to grip its head in the wounded branches in an attitude that was almost human."

The following day, on January 22, twenty days after the last German shell struck the Winter Palace of the Hermitage Museum, German field marshal Georg von Küchler flew to the Wolf's Lair to personally request permission to abandon Pushkin. The next day the Nazis fired their final shell at Leningrad. The Germans were by now retreating so quickly that Soviet troops were struggling to keep up.

Two days later the Red Army liberated Pushkin and Pavlovsk, where two and a half years earlier Kameraz had gathered up his precious potatoes while the Germans overtook the town's palace. The Institute's two experimental stations lay in plundered ruin.

At 8:00 p.m. on January 27, 1944, the surviving botanists from the Plant Institute gathered with thousands of others in parks and on bridges and embankments for the official victory salute. It would be another eight months before the battle for Leningrad officially concluded, after Soviet troops ousted the last German soldier from the country. But for the city's residents that night marked the end of the ordeal.

To commemorate, 324 guns situated across the city fired twenty-four salvos. Flames spat from their long barrels like trails from a lit firework. Red, green, and blue flares suspended by parachutes lilted down through the frigid air, lighting up the ravaged city and its ruined suburbs.

"The whole sky was lit by a phosphorescent glow as if a meteor had just flown past us," wrote Vera Inber. "First there was a plethora of crimson lights and then gold stars would stream downward like ears of grain from some bounteous invisible sack."

Searchlights from the Baltic Fleet moored in the river cast their beams into the sky, no more in search of enemy planes, but rather as a signal of liberation from the demonic forces of human warfare. In every meaningful sense, the siege—the longest endured by any city in recorded history—was finished.

That night, as she sat down to write her diary entry for the day, Inber fumbled for words that could capture and honor the weight of the moment.

"Though I'm a professional writer, words fail me," she finally wrote. "I simply state: Leningrad is free."

Four days later Anna Zelenova, curator of the museum in Pavlovsk, returned to the town. No trains or trucks were yet running, so she set out on foot. Clouds of crows swirled overhead, drawn to the lifeless figures of fallen German soldiers strewn unceremoniously by the roadside. One body had been propped up against a fence next to a note that read:

"Wanted to get to Leningrad. Didn't make it."

At Lannach Castle, a British POW and botanist, William Denton Venables, helped Heinz Brücher plant seeds stolen from Vavilov's collection in the surrounding hills at the behest of Heinrich Himmler. They were joined in their efforts by nine Jehovah's Witnesses from Ravensbrück concentration camp, who worked the land but refused to pull up weeds on religious grounds.

XVII

CASTLE OF SPOILS

I

FROM THE WINDOWS OF THE train bound for Leningrad, the passengers saw a world upended. Two decades earlier Vavilov and his young team, full of hope and vision, had made this journey through pristine snowdrifts. Now Vavilov was dead. And the evacuated botanists returned to Leningrad through an eerie tableau of desolation and ruin.

Bologoye, Lyubam, Tosno—each village a thatch of broken beams and pounded masonry. The driver slowed to cross the temporary, skeletal bridges erected to enable trains to run again from Moscow to Leningrad, allowing his passengers to bear witness to the harrowing aftermath of war. Wreckage remained uncleared; bridge supports cluttered the rivers like discarded matchsticks; everywhere signs written in both German and Russian warned of unexploded mines, those soldiers who never sleep. In places new telegraph poles, their wood as white as candles used for a somber ceremony, replaced those that had been burned or cut down.

It had been several years and a lifetime's worth of anguish since Sergei Vavilov had visited Leningrad. The war rumbled on in the west—Odesa had been liberated on April 10, 1944; Crimea on May 9—but four months after the siege broke, Sergei disembarked into a quiet, humbled city. More than 150,000 heavy artillery shells and 10,000 bombs and incendiaries had landed on streets, monuments, and buildings, smashing windows, cracking walls, punching holes in roofs, destroying entire frontages to expose the domestic cavities and warrens inside. Close to one in every three of the city's pre-siege population had perished. Most evacuees were yet to return.

Compared to their prewar bustle, Leningrad's streets appeared listless and abandoned. In one children's home, the news of the German retreat had been met with shrieks of happiness and flung pillows; then one child began to cry, then another, until every youngster was inconsolably sobbing. The joy of liberation quickly made way for grief as people finally allowed themselves to acknowledge their losses.

"How should we live now?" asked one survivor. "For what purpose?"

The failure of Soviet leadership to acknowledge the events of the past three years compounded the sense of purgatorial muddle. Soon after the blockade was lifted, Major General Ivan Fedyuninsky had informed Stalin of what the people had endured. In the constant barrage several thousands of lives had been lost to shelling by the Germans, but at least 649,000 had succumbed to starvation.

"The worst thing," Fedyuninsky told his leader, "was that someone dying of hunger would retain their consciousness through to the very end." It was as if they were watching the approach of their own death. "The siege of Leningrad," he concluded, was "one of the great tragedies in human history."

Stalin, eager to forestall any accusations that his government had failed its people, retorted that their suffering was not, in fact, unique. "Death did not only cut down Leningraders," he said. "There was nothing more we could have done for the city. . . . War and death are inseparable. Leningrad was not the only place to suffer."

The Plant Institute was just one of hundreds of enterprises to suffer during the blockade. And yet, burdened by the knowledge of the death of his brother, when Sergei Vavilov arrived at its heavy wooden doors, he felt an unusual sense of loss and abandonment. Here, he knew, he would be faced with a colossal absence.

The building was almost empty except for a few remaining workers who had not starved to death, perished in artillery attacks or bombings, or fled across the frozen expanse of Lake Ladoga. The seed bank that represented his brother's life's work, the countless miles of exploration and careful specimen gathering, had been all but abandoned.

The seeds, too, had mostly gone. Shortly after the siege broke, a group of scientists had arrived at the Institute from Krasnoufimsk in the Ural Mountains, where some of the collection had already successfully been evacuated by train and plane. When they arrived, they had packaged up most of the remaining collection, including seeds and potatoes that had been renewed by Ivanov's team at the Predportovy collective urban farm, in the south, and Lekhnovich and Voskresenskaya's team at the Lesnoye collective farm, to the north. The scientists then transported the crates through the city and onto the trains that would carry them more than a thousand miles east, into the Urals for safekeeping.

The world collection had been saved, albeit at a harrowing cost. Stripped of its seeds as well as its staff, a fog of impenetrable sadness seemed to have settled inside the building. That night in his diary, Sergei wrote, "If only God and souls could be brought back inside these walls."

In a spirit of melancholy Sergei left the Institute and walked through streets dressed in the colors of an early summertime evening. The emptiness of the once-bustling Nevsky Prospekt seemed to mirror his inner turmoil: the sharp sense of loss, of guilt about the past, and deep concern for the future. In the distance, a solitary light signaled from the window of his late brother's former apartment. As he neared, Sergei could see it was a simple lamp; its orange shade cast a gentle glow upon the room within, where somewhere its new occupant might have been brewing tea or reading a book or whatever it was that ballerinas did in their spare time. To Sergei, it represented a poignant symbol of both profound loss—the stolen legacy of his brother, ruined by the duplicitous accusations of an oppressive regime—and the ever-changing doggedness of life.

II

THE GREAT PATRIOTIC WAR—A WAR within a world war—was not yet finished, but Leningrad had already begun to commemorate and valorize its survivors and its dead. In the summer of 1944, an exhibition opened in the cool expanse of the Solyanoy Gorodok building, a short walk from the Plant Institute, on the bank of the Fontanka River. Children clambered atop the captured enemy artillery guns that stood to attention in front of the entrance opposite the summer garden. Inside, the exhibition spread through various halls and trickled down into the basement.

In the main hall a pyramid of German helmets stood so tall it almost touched the ceiling. Some of the 154 mm guns used to shell Leningrad's people stood beside six-barreled flamethrowers and a showroom of battered tanks painted in green, granite, or snow white. Cards bearing museum-style descriptions informed visitors of the size and weight of the unexploded munitions on display, along with statistical readouts of the number of shells and bombs that had dropped on the city on specific nights.

Another section recorded the feats of Leningrad's industrial workers: the number of turbine generators produced, or the output of tobacco and yogurt factories that had been redeployed to send arms and munitions for the defense of Moscow. There was mention of the ingredients used to make blockade food: the root flour and sweepings kneaded into scones; the dextrin that had been turned into fritters; the cellulose used for baked puddings; the fish glue used for jellies; the pigskin belts that had flavored water just enough that it might be described as soup.

The visitors who toured the exhibition still bore the signs of malnourishment: the pale, yellowish skin known as the blockade pallor. In one area sat a humble set of weighing scales. One side of the scales bore four small weights that together measured four ounces, a reminder of the unsustaining sustenance provided to most individuals during the first terrible siege winter. Nowhere was there mention of the sacrifice of Vavilov's botanists, those who had given their lives to protect the seed collection, or the survivors who had deployed their skills to help feed the population.

III

A FEW WEEKS LATER, IN the autumn of 1944, Red Army soldiers led the first group of German prisoners of war into Leningrad. The POWs were brought to repair and rebuild the buildings that they and their comrades had relentlessly shelled: a "you broke it; you fix it" policy.

"As if by some invisible sign, people emerged from nowhere to watch," recorded one eleven-year-old who saw the German soldiers arrive. "They stood silently at the pavement's edge, close to the prisoners, unable to tear their eyes away." In the spreading sunshine the survivors of the blockade stood and watched the prisoners traipse by, not looking up. "Their faces were exhausted and tense," she noted. "Some looked straight ahead, some walked with their heads lowered; a few tried speaking with each other, pretending that they were paying no attention to us. We all stood silently. There never was a single shout, curse, or insult. Not a sign of anger or hatred. Nobody addressed them at all. We stood like a thin immobile wall and behind us stood the ghosts of our dead."

There was much practical repair work. For its restoration the Hermitage needed sixty-five tons of plaster, a hundred tons of cement, six thousand square yards of glass, eighty tons of alabaster, and thirteen pounds of gold leaf. At the Plant Institute both buildings had been damaged by incendiary fire, with the archive and herbarium sustaining the gravest harm; the cost of damage was estimated at 14.4 million rubles—several hundred thousand dollars.

For the few remaining staff members, the gravest concern was for the seeds. Buildings could be repaired, library books returned. Many of the collection's seeds, however, were irreplaceable, their original habitats lost, or now, in the chaos of a world war, unreachable. An unfathomed portion was now under Nazi control, watched over by a British solider-botanist and his SS captor, who, in the face of the advancing Red Army, had just received an instruction to blow up the castle that housed the stolen seeds.

IV

SINCE ARRIVING AT LANNACH CASTLE, William Denton Venables had been joined by a ragtag band of fellow gardeners. In the spring of 1944, a group of Jehovah's Witnesses had arrived, most from Ravensbrück concentration camp. These nine women, whose average age was forty-four, originated from Germany, Bohemia, Moravia, and Poland. They had changed their camp clothes on arrival at the castle, but continued to wear the purple triangle used by the Nazis to identify them from other classes of prisoners, the Roma, the homosexuals, the dissidents, and the Jews.

The women had been chosen for the task because of their faith. Jehovah's Witnesses never tried to flee. "Properly used, you can leave them in the fields without supervision, they will never try to escape," Himmler wrote in January 1943 to Oswald Pohl, head of the SS's main economic and administration office. "If you give them tasks to solve on their own, they will become the best administrators and workers." The refusal of Jehovah's Witnesses to touch weapons or engage in munitions work rendered them unsuitable for several vital industries. But in the fields these believers could find purpose.

At Lannach the women assumed diverse roles. Every morning a guard lined them up in pairs, then led them to their designated stations. Some deployed to the SS test laboratory, others to the greenhouses. The remainder were assigned tasks in the fields, where they sowed, planted, and harvested. The women occasionally worked side by side with civilian laborers or found themselves among fellow forced laborers, tilling the soil and tending to the land. They worked diligently, only clashing with Brücher's team about weeds. Their faith, they explained, forbade them from pulling any living thing from the soil. Theological disagreements aside, the women proved compliant. When it became clear none would leave the castle or its grounds, the SS warden charged with guarding them, Emma Raabe, was redeployed.

With the German withdrawal from the Eastern Front fully underway, the Wehrmacht had been eager to return Brücher to his artillery role, where he might help slow and even reverse the retreat. Brücher wanted to remain in Lannach, however, where he might work with the treasures his commando team had collected from the Soviet field stations. In February 1944, Brücher's request was granted; he was formally and permanently transferred to the Waffen-SS, awarded the rank of SS *Untersturmführer*, and administratively attached to Himmler's personal staff.

Himmler was keenly interested in farming, which he saw as intricately linked to Nazi efforts to raise a master race (before the outbreak of war he had become obsessed with trying to breed pure white chickens). He likened his role hiring for the SS to that of "a nursery gardener trying to reproduce a good old strain that has been adulterated and debased; we started from the principles of plant selection and then proceeded quite unashamedly to weed out the men whom we did not think we could use." Himmler's interest in Brücher and his ragtag team's efforts was detailed. One particular experiment caught his eye: an endeavor involving a South American oil plant called Chilean tarweed—*Madia sativa*. This resilient plant, capable of yielding a valuable cold-resistant oil, could prove useful to the war effort. And should it prove fruitful, Himmler reserved the right to christen this botanical triumph.

The tides of war soon denied the request. Nine months after Sergei Vavilov visited the forsaken seed bank in Leningrad, Brücher received an instruction from the Ahnenerbe's secretary general, Wolfram Sievers, to blow up the castle the moment the Russians advanced on Lannach. If the war was lost, the seeds, at least, would not return into enemy hands.

Brücher, however, was unwilling to destroy the parts of Vavilov's collection that he had dedicated the past two years to securing. Not wanting to contradict a direct order, he developed a ruse that might save the specimens, while ensuring they would not fall into enemy hands.

"Some of the Tibetan seeds I will sow in the mountains this spring," he wrote in response to Sievers. Brücher planned to hide the specimens in plain sight, in fields and gardens away from the castle grounds, where only an expert would be able to discern that their stems and leaves belonged to varieties not typically found in the region.

Brücher asked Venables, who had access to a shortwave radio, to keep him abreast of developments on the front. The Englishman seemed to have better intelligence than his captors about the state of events on the outside.

"I have analyzed and prepared for any situation that might arise as the front approaches," Brücher assured Sievers, admitting to his superior that he had identified safekeeps in the mountains where he might stow "valuable seeds, important appliances, and tools" at a moment's notice.

V

IN EARLY APRIL 1945, JUST as the prisoner botanists began to plant out the second year of seeds in the castle grounds and surrounding mountains, Venables decided the time had come to stage another escape attempt. Via broadcasts from London to Austria he had learned that the arrival of the Russians was imminent. So, one evening, while idling in one of the castle's larger rooms used to house farm-working parties, he asked for permission to visit the latrine.

Instead, Venables made for the woods. By now he knew the local countryside well; it would be far easier to remain hidden now than during his first escape attempt in Greece. In his pocket he held a list handed to him by another British POW of locals who were in sympathy with the Allies. For the next few days, the Englishman tentatively visited the addresses listed and met with supportive Austrians, who provided him with food and blankets.

From his vantage in the woodland, Venables spied the advancing Red Army troops. On May 4, 1945, he witnessed the German army begin to stage a "disorganized withdrawal," as he described it. He watched troops cut horses loose and discard and destroy equipment—tanks, guns, cars, and trucks—in what he believed to be an effort to reduce traffic on the roads out of the area. Panic set in. Venables was eager to ensure Vavilov's seeds were not destroyed, either, by the fleeing Germans or the advancing Red Army soldiers. He had another motive. When he returned to England, he planned to establish a plant nursery. Perhaps he might salvage some of this rare plant material for his own benefit. He just needed a reason to return to Lannach Castle, which was still overrun with SS officers and guards.

In the early hours of May 8, Venables put on his uniform, found an abandoned German staff car by the side of the road, and drove into the town of Graz to present himself at the Russian headquarters, recently established in the town. There, he inquired as to the plan for British POWs in the area. The Russians explained that they were willing to transport any Allied POWs in Russian-occupied territories twenty-five miles toward the common frontier.

With the necessary pretext Venables thanked the Russian officers and set out for the castle planning to collect his comrades still held in the building and, perhaps, some of the collection. As he drove toward Lannach, Venables spotted some weapons laying discarded at the side of the road. He pulled over and loaded the guns into the vehicle, then continued to the castle. He pulled up to Lannach and, weapon in hand, entered the building, shouting for any occupiers to offer their surrender.

"Immediately the staff, including SS officers and men, surrendered the whole establishment," he later recalled. Not wanting to take prisoners, or perhaps not wanting any more witnesses than was necessary, Venables told the Germans that the war was over and they were free to leave. Brücher fled. Then Venables sent the British POWs who had been held in the castle out to inform all other friendly prisoners in the area to report to Lannach, from where they would be safely escorted to Allied-held territory.

Soon around 150 French and British soldiers had amassed at Lannach. Venables oversaw the distribution of provisions from the castle's stores and, when food stocks began to run low, set out to "collect what was required" from the surrounding area.

Russians began to visit the castle every day. Via a Polish interpreter, Venables told the visitors that they could occupy the castle as soon as he and his fellow POWs had moved out. Six days later, the Russians arrived again, this time insistent that they needed to move in. Venables headed for the frontier, holding a chit written by his interpreter granting him permission to leave and signed with a fake Russian name.

On May 18, 1945, laden with bags, Venables crossed the frontier and began his return journey to England. For now, the whereabouts of the stolen collection would remain a mystery.

VI

TWO MONTHS LATER, IN JULY 1945, Abram Kameraz returned from his posting in Latvia to a city, a workplace, and a home irretrievably changed by war. Close to three million Soviet servicemen taken prisoner during the war had died from starvation or disease or had been accused of desertion and shot by the Soviet state. Tens of thousands more, having been liberated by their fellow countrymen, were now sent to the gulag. Those like Kameraz fortunate enough to return home to Leningrad faced their own challenging reunions.

Everyone had been distorted by war, physically and psychologically. The old attachments between parents and children, wives and husbands, no longer fitted in familiar ways. In some homes, returnees discovered new occupants, who, having taken up residence in what they believed to be abandoned properties, were unwilling to move out. A law that required the return of valuables that had been sold at bargain prices during the most trying months of the siege was poorly enforced. Everywhere there was evidence of loss.

After three and a half years at the front, Kameraz had been promoted to the rank of sergeant major, sustained two injuries, and amassed a cabinet's worth of medals: the Order of the Red Star, the Valor medal, the Defense of Leningrad medal, the medal for the Victory against Nazi Germany, and the Excellent Reconnaissance Officer badge. No amount of decoration could compensate for the deeper losses he had sustained. The potatoes Kameraz had carried on his back had survived the siege. His wife, Lyudmila, who had joined her husband in guarding the crop, had not. Nor had the couple's baby.

Kameraz returned to the Institute. He was appointed senior research officer in the root crop section. In the seed bank the losses of war were keenly felt and enduring. The slow work of restoring the collection, of removing the tins from their hiding places, and relabeling and reclassifying the surviving seeds was colored by the absence of lost friends and colleagues, those who had given their lives to save their work, and of course their leader, murdered by a duplicitous state. Consequences rippled. On March 3, 1949, Lekhnovich's wife, Olga Voskresenskaya, died from complications related to the injuries sustained from a shell blast while planting potatoes in the fields during the siege.

VII

IN JULY 1945 STALIN PROMOTED Vavilov's younger brother, Sergei, to the position of president of the USSR Academy of Sciences. While Sergei was eminently qualified for the role, Stalin had only become aware of the true extent of Nikolai Vavilov's international reputation after the British prime minister Winston Churchill inquired about the Soviet botanist during the Tehran Conference. Stalin was now eager to deflect blame for the botanist's disappearance and death.

Sergei immediately used his newfound position and power to help his family. A few days after his appointment Nikolai Vavilov's younger son, Yuri, received a telegram summoning him and his mother, Yelena, to Leningrad, where Yuri was to assume a position at the Academy of Sciences. He was a seventeen-year-old high school student; there was no job for him, but the ploy enabled the family to leave Saratov for Leningrad.

En route, the pair first visited Sergei at a Moscow hotel. There he explained that Yuri and his mother would be unable to return to the apartment they had lived in with Vavilov before the war, but that Sergei had secured for them a small apartment within the academy's buildings on Vasilyevsky Island. The following summer Yuri graduated from high school with a silver medal, an award that enabled him to enter Leningrad University without taking any further exams.

Sergei was unable, however, to save his brother's elder son, Oleg. Two and a half years after he visited Saratov and there learned about his father's fate, he earned his PhD at the Lebedev Physical Institute on December 20, 1945. To celebrate his doctorate, on January 24, 1946, Oleg took a train with a group of other graduate students and professors to the ski resort of Dombay in the northern Caucasus. His young wife, Lidia, pleaded with him not to go, but Oleg told her he needed a chance to unwind, and besides, there were some interesting people among the nine-person party, such as the future star mathematician Igor Shafarevich.

On January 29 Oleg and his fellow travelers reached Nauka, the base camp in Dombay. At the last minute they had been joined by another climber, Boris Ivanovich Schneider, a forty-one-year-old graduate student who claimed to have conquered twenty-seven peaks. At the last minute, Schneider suggested the group take a different route to that which they had planned and climb Mount Semenov-bashi.

Oleg and his friends did not have the necessary equipment for this demanding climb, but Schneider insisted that, as experienced mountaineers, they would be fine. After some back-and-forth, the group agreed to the plan and on February 3 began the ascent. That night they slept in a Zaporozhian Cossacks camp, and on the morning of the fourth Schneider convinced everyone except Oleg to remain on the slopes to enjoy some skiing. Then he led Oleg on an unusual route toward the only deadly section of the mountain, with drops the height of a six-story building.

As the pair inched along the cliff face, Schneider announced that he had mislaid his ice axe and needed to untie the safety ropes from the rock face so he might search for it more easily. Oleg remained on the cliff face, in boots unsuitable for such perilous terrain, and without a safety harness. He did not hear Schneider approach him from behind until it was too late; Schneider struck Oleg in the right temple with the axe, a blow that caused Vavilov's son to lose his footing and tumble into the abyss below.

Schneider descended the cliff face to check and hide the body. If he had been sent to assassinate Vavilov's son at the behest of the secret service as a measure to tie up loose ends, then his mission was complete. At four in the afternoon, just as the mountain was falling into darkness, he returned to the camp and told the other members of the group that there had been an accident. Oleg was missing, presumed dead.

Seventeen days later Oleg's wife, Lidia, and a group of his friends traveled to Dombay to search for her missing husband. His murderer, Schneider, accompanied the group, who needed him to take them to the spot where Oleg fell. At first Schneider attempted to excuse himself from the search, but the group insisted. When they arrived at the scene, Oleg's friends grew increasingly angry and accused Schneider of Oleg's murder.

Unable to find a body, the group returned to Moscow. In mid-May Lidia returned to the mountain. Each day she set out to comb the ice, which had begun to melt, for signs of her husband. On June 10, a few days before she was scheduled to return home, Lidia caught a flash of some red clothing under the ice. She began to uncover the body of her husband, perfectly preserved in the ice, his cheeks still flushed red. Oleg's death certificate noted a "wound in the area of the right temple the size of an ice pick's adze."

His family did not doubt the cause of death. Oleg knew too much about his father's death. If his ascent in Soviet science continued, Vavilov's story might come out.

VIII

IN THE HALLS OF THE Institute, Ivanov and the bereaved Lekhnovich were left as fixtures, both as consultants and as symbols of a bygone era. The precise fate of their leader and his eldest son would, for now, remain a mystery. Lekhnovich, who had grown a long white beard that he combed into a cascade that fell to his waist, appeared as though he had weathered countless winters. Still, he refused to wield seniority over younger staff at the Institute, waiting in line at the cafeteria when, as a department head, he had the right to skip the queue. His wartime experience had made him conscientious and thrifty: he kept a pile of scrap paper, using the blank sides to draft reports, and used the same, increasingly worn chair for his entire career. After he remarried, the couple named their daughter Olga, in memory of his first wife.

Ivanov assumed the weighty responsibility of research secretary to the Academy of Sciences committee dedicated to preserving Vavilov's legacy. He approached his role with the fervor of a heartbroken archivist, gathering a collection of more than four hundred photographs of his friend. Ivanov, who suffered from hepatitis throughout the remainder of his life, worked quietly but determinedly for the restoration of Vavilov's reputation, as well as that of the seed bank staff who had endured so much.

Foreign inquiries concerning Vavilov's whereabouts had gone unanswered by the Soviet authorities. In June 1945, the month the United States, the USSR, Britain, and France signed the Berlin Declaration, confirming the complete legal dissolution of Nazi Germany, scores of foreign scientists attended celebrations held in Moscow to mark the 220th anniversary of the Academy of Sciences. There, members of the British delegation finally learned that Vavilov had died in unknown circumstances, probably in the prison at Saratov.

When they returned to Britain, Vavilov's English scientist friends Sydney Harland and Cyril Darlington cowrote a mournful, affectionate obituary, which was published in the November issue of the journal *Nature*. After outlining Vavilov's accomplishments ("a remarkable sight to see [him] work") and character ("vigorous, confident and cheerful"), Harland and Darlington turned their attention to the fate of the collection Vavilov had amassed.

"When Leningrad came to be besieged," they wrote, "the residue of his collections was eaten by the famished people."

When Ivanov and his surviving colleagues read the obituary, this dismissive aside both wounded and infuriated them. To see their efforts and those of their perished friends misrepresented in such a prestigious international publication was indescribably painful. Having suffered so greatly and lost so much from their commitment to Vavilov's vision and their determination to protect the seed bank, no matter the cost, here was yet another betrayal of truth.

Ivanov issued an invitation for Darlington to come to Leningrad and see the evidence for himself. When Darlington arrived, Ivanov led the Englishman through the dark corridors and into the rooms to which, by now, the collection had returned. On the shelves sat rows of tins that a few years earlier the staff had bound together, away from ravenous vermin and their own temptations.

"We showed him that the collection was saved almost to the last seed," Ivanov recalled.

Embarrassed, Darlington explained how he had learned that the staff ate the collection from a BBC broadcast. He apologized for repeating the error. Nobody in England, he explained, could have believed the collection had survived while its custodians and their city starved.

Nikolai Ivanov (*top row, second from right*) with workers from the Plant Institute
after the end of the siege. By 1967 a hundred million acres of Russian agricultural
land had been planted with seeds derived from the collection saved by the
scientists' sacrifice. By 1979 that area had almost doubled to a third of all Russia's
arable land. Today, the seed bank in St. Petersburg holds more than 320,000
separate samples, with more than 4,500 new and unique types of plant bred from
original samples collected by Vavilov and his teams.

POSTSCRIPT

I

OF THE QUARTER OF A million accessions held at the Plant Institute buildings in Leningrad at the start of the siege, around forty thousand—two metric tonnes' worth—were consumed by vermin or failed to germinate after the war. In addition to the seeds taken by Heinz Brücher, the retreating German army captured or destroyed the subcollection at the Detskoye Selo laboratory in Pushkin, including ten thousand accessions of wheat, barley, oats, vegetable crops, scientific equipment and library books, and a collection of sixty-six thousand varieties of flowers gathered by Vavilov during his expeditions around the world.

These losses extended to the Institute's staff. In addition to those who died on Leningrad's streets or at the front line, at least nineteen staff died inside the Plant Institute buildings during the siege, most to starvation, some inside their freezing offices while surrounded by containers of edible plant matter that, if consumed, might have extended or saved their lives or those of the people around them. Through collective accountability and encouragement these men and women chose to abstain despite an instruction from the Institute's director to "spare nothing to save the people."

Their deadly decision was rooted in two core beliefs: that many of the samples were irreplaceable due to the loss of natural habitats from which they were first collected, and that there could be hitherto unrecognized genetic qualities in these wild varieties of plant that would in time prove important to plant breeders. The survivors were soon proven correct. Varieties of wheat collected by Vavilov from Spain, Japan, Italy, and Argentina and saved by the staff were crossbred to create the winter variety Bezostaya 1, in time used across the world for its high yield.

Samples of a rare and disease-resistant variety of wheat—*Triticum persicum*—collected by Vavilov in the mountain valleys of Dagestan were used by British and Australian plant breeders to develop a new, high-yielding variety. Kameraz, who saved the potatoes from occupied Pavlovsk, developed dozens of new varieties, including Detskoselsky, a three-species hybrid resistant to viral infections based on one of the rescued Bolivian plants, which became a favored crop of farmers in the region for decades.

By 1967 a hundred million acres of Russian agricultural land had been planted with seeds derived from the Institute's collection. By 1979 that area had almost doubled to a third of all Russia's arable land. Today the seed bank in St. Petersburg holds more than 320,000 separate samples, a collection that has proven invaluable in ensuring food security in Russia, with more than 4,500 new and unique types of plant bred from original samples collected by Vavilov and his teams.

Ninety percent of the seeds and planted crops held in the St. Petersburg collection are found in no other scientific collections in the world. More than a thousand crossbred varieties of these plants bear the name of their discoverer, Vavilov. In kitchens and dining rooms around the world, people continue to benefit from the sacrifice of the men and women who gave their lives during the siege.

But while the surviving botanists were undisputed heroes of science, their story was beset with political complications. As the fruits of their work and sacrifice spread around the world, their story remained untold.

II

ON AVERAGE, ONE BOOK ABOUT the Great Patriotic War was published in the Soviet Union every single day between 1945 and 1991. Given forceful censorship and classified documents, most of these works had a homogenous feel. None but the most pristinely heroic memories survived the state's withering gaze to endure on the page. Some truths endured in euphemism; most were discarded in favor of palatable clichés. Published biographies were invariably mythopoeic, promoting a vision of a city, Leningrad, united in noble, selfless suffering.

Tales of siege heroism and sacrifice abounded, laying the foundations of a collective hagiography of residents who prioritized one another even as they perished. Many of these stirring accounts of selflessness and altruism were unrealistic, but, seemingly, a necessary part of grieving for a nation that wanted to honor those whom its government failed to save.

The Soviet state was especially incentivized to cleave the heroism of the city's defenders from the context that necessitated that heroism. The blockade was initiated and enforced by the German army, but the mass death event was closely connected to homeland decisions: the choice to remove food lest it fall into enemy hands; to store rations in concentrated areas vulnerable to bombing; to terrorize with capricious arrests made on spurious grounds; to impose unnecessary inequalities on the populace as ordinary people saved up their crumbs while the corrupt bureaucrats banqueted.[*]

To the state, private memory was risky. Any individual whose diary, memoir, poem, or novel offered a more nuanced recollection of life during the blockade was subject to censorship on the grounds that complicating the story would offend the memory of those who died that the recaller might live. Leningrad literature was, as the scholar Eugene Ostashevsky put it, "two-tongued duplicitous literature," a body of work with a hidden basement containing unseen depths.

This is the context into which the scientists who survived at the Plant Institute emerged. It is almost impossible to comprehend the environment in which Vavilov and other Soviet scientists worked during Stalin's reign of terror: the strain of trying to understand the hard realities of the natural world beneath a regime unwearyingly committed to breaking the truth. Vavilov was gone, his whereabouts unknown. His closest friends and most senior staff had also vanished; many were dead.

---

[*] Some historians argue that, if not for these imposed inequalities, none would have died in the city from hunger.

The regime had failed the seed bank. The lack of clear, early, and decisive action meant the botanists were left floundering, unsure of what to do, even while the treasures of the Hermitage Museum across the square flowed to safety. Abandoned, neglected, and leaderless, those left behind to steward the collection had only their own moral and ethical standards to guide their behavior and policy, even while, person by person, they began to starve. That their leader had been branded an "enemy of the people" and arrested further isolated the Institute. The risk of sharing a story that was incompatible with the state's purposes was clear to every surviving botanist. It would be years before fragments of truth began to emerge.

III

AT FIRST THE SOVIET UNION was keen to forget. Until the death of Stalin in 1953, Leningrad's leaders erected no memorial to those who perished during the siege. Anti-begging decrees banished disabled veterans to the islands of Valaam in the far north of Lake Ladoga. Only in 1956, after the Soviet leader Nikita Khrushchev denounced Stalin and decreed a new era of openness and international cooperation, did workers begin to remove the fences and clear the nettles that had overrun the mass graves at Piskaryovskoye and elsewhere. At last, those traumatized by the siege were afforded the chance to remember and grieve.

That year Nikolai Vavilov, who had been expelled from the Academy of Sciences in 1940 after his arrest, was posthumously reinstated as an academician. His brother, Sergei, who had died of a heart attack in 1951, never lived to see the restoration of his brother's reputation. Two years later, in 1958, Dmitri Pavlov, chief of food supplies during the siege, published a personal and comparatively candid account of the besieged city that would have been impossible to deliver during Stalin's lifetime. In the book, Pavlov mentioned "an institute of plant genetics" that had "dropped from sight in the commotion of war. The authorities had no time for it."

Pavlov outlined the basic story and named a few of the employees who had died of starvation,* as well as the survivors who "continued with the project of saving the collection at any cost." For Pavlov, the story demonstrated how a group of citizens "triumphed over their suffering" with stoicism and pride, an episode indicative of the character of the Soviet people, revealing a "love for the society they have created," one that "guided their struggles with the invaders, hunger, and other deprivations." Here was a harmonious story within a story: a group of faithful citizens whose choices exemplified the broader narrative.

The botanists' story was already enshrined in the language and structure of heroism when it was announced to the world: a group of people whose valor lay in their capacity for endurance and self-control, their willingness to preserve life beyond the boundaries, both physical and temporal, of the building and its context. They chose death over self-preservation, a choice made in service of a greater good: a condensed parable for a siege that had already begun to calcify into myth.

Vavilov, however, complicated what might otherwise have been a straightforward tale of wartime courage. The scientists gave their lives to save a collection that the Soviet state had abandoned, amassed by a botanist it had murdered. This secret poisoned the legend. The true circumstances of Vavilov's murder now needed to be brought to light.

IV

IN APRIL 1965 THE UKRAINIAN writer and academic Mark Popovsky secured access to ten thick files that referred to case number 1,500 at the offices of the procurator general in Moscow. Popovsky had in the fifties begun to write a biography of Nikolai Vavilov, intending to discover the truth of his fate. (A fellow professor told Popovsky at the time, "There must be a simpler way of killing oneself.")

With access to the documents, Popovsky became the first outsider to learn the story of how and why the security services arrested Vavilov, who informed on him, what sentence he received, and how and where he died. Popovsky copied notes into seven blue school-exercise books,

---

* See "Staff Roll Call" on page 309.

recording in detail how Lieutenant Aleksei Khvat had subjected Vavilov to innumerable interrogations until the botanist agreed to collude in the fictional version of events that had been prepared for his confession. Later that year Popovsky began to present lectures in which he referred to Vavilov's persecution. One of these lectures, delivered at the seed bank itself, had a particularly dramatic effect.

"I saw people in tears, and informers whom I mentioned by name jumped up and left the hall to the hissing and jeers of their colleagues," he recalled.

The following year Popovsky published an article in the provincial magazine *Prostor*, in which he described the three years leading up to Vavilov's arrest, and Trofim Lysenko's role in these events. This mild provocation was enough to move the state to issue a two-year ban on the publication of Popovsky's writing.

In 1967, prior to the eightieth anniversary of Vavilov's birth, the Soviet Academy of Sciences granted Popovsky and Professor Fatikh Bakhteyev, one of the last employees at the Plant Institute to see Vavilov alive, permission to travel to Saratov Prison. Popovsky had already written to medical staff who had served in the prison hospital ward asking for details about Vavilov's final days, but none could remember or was willing to remember the circumstances of their patient's death.

In Saratov, Popovsky was further confounded: the local police force had destroyed Vavilov's case file when the mandatory period for preservation had expired. After several days in Saratov, Popovsky and Bakhteyev received a summons to the office of Colonel A. M. Gvozdev, who passed them a few faded pages from the prison records detailing the circumstances of Vavilov's death, listed in the nurse's report as "dystrophy"—the same affliction which had killed more than a dozen of Vavilov's colleagues in the seed bank.

By 1967 attitudes toward Vavilov and his rival Lysenko had sufficiently shifted that the Institute could adopt Vavilov's name as its own, becoming the N. I. Vavilov All-Russian Institute of Plant Genetic Resources. In December 1968 Moscow named a celestial landmark after Vavilov and his brother, Sergei. "The rehabilitation of the great Russian geneticist has been extended as far as the moon," wrote one journalist for the *New York Times* in a story titled "Moon Crater Named for Once-Disgraced Soviet Geneticist and Brother."

While publicly engaged in the restoration of Vavilov's reputation, the Soviet Union sought to suppress details about his demise. The same year Vavilov arrived on the moon, a KGB agent visited Popovsky's apartment in Moscow. The agent issued a stern warning to the writer not to communicate, in conversation, lecture, or publication, the information he had obtained from Vavilov's case files.

The visit did not dissuade Popovsky from his work, which had by now acquired for the writer and historian a new sense of urgency and mission: as the only person to have been granted access to Vavilov's security files, he alone had proof of the persecution the botanist experienced at the hand of the state. Popovsky continued his work in secret, interviewing more than a hundred of Vavilov's coworkers, students, and relatives.

The following year the *New York Times* journalist Harrison Salisbury published his seminal work on the siege, best known by its later title, *The 900 Days*, which described in detail the suffering of Leningrad's people and the associated failures of their leaders. Salisbury's history included no mention of the Plant Institute, however. In Russia that same year, Ivanov, Kameraz, and Lekhnovich each contributed a brief chapter to a state-published history, *V Osazhdennom Leningrade* (In the Besieged Leningrad). The men described, with obvious restraint and omission, flashes of their experience.

While enduring the ban on his work in the Soviet Union, Popovsky continued to work on his biography in secret, which included a report of his visit to Saratov to locate the site of Vavilov's grave with Professor Bakhteyev. Popovsky began to disseminate hand-printed copies of the first draft among "intellectual circles in Moscow and Leningrad," as he described them. He gave copies to trusted friends to smuggle out of Russia and take to Central Asia, Kyiv, Voronezh, and the Far East. Thousands began to read his unsigned account, which was, according to a report in the London *Times*, "in such demand as to be almost unobtainable" in Russia's capital.

On June 3, 1977, the KGB searched Popovsky's apartment for four hours, looking for the notebooks into which the author had copied out details contained in the clandestine files. They did not find the notes. (When the officer, Captain Bogachev, picked up the old leatherbound file containing the notes and asked Popovsky what it contained, the writer replied nonchalantly, "Wastepaper.")

Even if the authorities had discovered the exercise books, Popovsky had already photographed his notes and distributed them among friends and colleagues both within and outside Russia. These persistent acts of journalistic courage salvaged from oblivion the story of Vavilov's fate and exposed the culture of fear and paranoia that led to his betrayal by contemporaries, and subsequently his murder.

In 1979 Ivanov and some other surviving staff spoke to Viktor Senin, a journalist for the state-sanctioned newspaper *Pravda*, who wrote a short but compelling historical narrative of life inside the Institute during the siege—the only published book-length work on the subject in any language until now. Senin included scenes of death in the building and even described disagreements between staff members, some of whom argued that they should consume the seeds to preserve their lives. The book, which is captivating but clearly embellished, laid out the story, which again harmonized with the encompassing myth of Leningrad's collective, flawless sacrifice. "They believed in the future and won," Senin wrote in his conclusion.

In 1987, on the centenary of Vavilov's birth, conferences were held in Moscow, Leningrad, London, and the German city of Gatersleben. In his opening address for the symposium held by University College London and the Linnean Society—of which Vavilov was a Fellow—Vavilov's friend John Gregory Hawkes, who in 1938 had visited Leningrad to observe the Institute staff's work with potatoes, said of his friend, "Vavilov was probably the most distinguished plant breeder, agroecologist, and applied geneticist of his generation." In his shortened career, Hawkes continued, Vavilov "laid the foundations of modern plant breeding."

The centenary, as Popovsky put it in a letter to V. A. Korotich, editor in chief of the magazine *Ogonek*, made it possible to "untie the knot" of Vavilov's story and make the secret public among ordinary

Russians. Dozens of articles appeared in the mainstream press, many of which drew attention, to one degree or another, to Vavilov's fate.

Popovsky was dismayed to discover that much of his work had been quoted without credit. A. L. Takhtadgigian, the author of what Popovksy considered to be the "the richest in content and the frankest" of these articles, published in the Russian cultural newspaper *Literaturnaya Gazeta*, claimed that he had credited Popovksy in his original draft but that the editor had struck these references prior to publication without Takhtadgigian's knowledge.

In a letter addressed to the publication's chief editor, Alexander Chakovsky, Popovsky threatened to take legal action, writing, "You knew that I was the primary source of all the facts, quotes, and dates presented therein. . . . Due to political considerations you were ordered to wipe out my name because I am a writer-émigré." Demanding that Chakovsky publish his letter, Popovsky added, "I am convinced that the Soviet press, which has kept silent about Vavilov's murder for 45 years, will not be completely honest in this case either." Still, in 1987 Vavilov was commemorated on a Soviet stamp, a clear and public sign of approval, if not explicit remorse, by the state.

Four years later two staff members from the Institute cowrote a feature for *Diversity* magazine titled "Vavilov Institute Scientists Heroically Preserve World Plant Genetic Resources Collections during World War II Siege of Leningrad," which included photographs of the botanists, both those who lived and those who died. Their motive was, perhaps, to apply international pressure to the Soviet government to preserve the Institute, which was by now so desperately underfunded it was reportedly unable to pay its electricity bill. A few months later a journalist for the *Washington Post* reported the facts contained in the *Diversity* piece, and for the first time the basic story of the botanists' sacrifice during the siege of Leningrad became well-known in the international scientific community. (Although the inaccuracy reported by the BBC in 1945, that staff ate the collection, persisted. For example, Vavilov's entry in the 2002 *Cambridge Dictionary of Scientists*, second edition, states, "His seed collections were largely eaten during the siege of Leningrad.")

Today the story is familiar to many seed bankers, at least in outline and principle. In his 2007 *New Yorker* piece about the Svalbard Global Seed Vault in Norway, which contains a consignment of seeds from Vavilov's collection, the writer John Seabrook observed that "the story of what happened at the Vavilov Institute has a mythic resonance in the mind of every seed banker." This resonance is both intimate and provocative; it invites any keepers of seeds to ask themselves what they might do if faced with a comparable situation. Eat or abstain?

And yet, beneath the surface of a simple tale of choice and cost sit a cluster of more taboo questions: What sacrifices are permissible in the pursuit of scientific progress? At what point should a person choose to disobey a directive? What responsibility do we each have to the survival of future generations, and what is the correct course of action when that responsibility sits at odds with our obligation to the living?

V

IN THE RELATIVELY FEW INTERVIEWS that survivors from the Plant Institute gave, they spoke in broadly emotionless terms of how the moral, mortal dilemma they faced was, in fact, no dilemma at all. Each would sooner die than sacrifice Vavilov's collection. "It was hard to walk. . . . It was unbearably hard to get up every morning, move your hands and feet. . . . But it was not in the least difficult not to eat up the collection," said Lekhnovich. "It was impossible to eat it up, for what was involved was the cause of your life, the cause of your comrades' lives." Vavilov's favorite student, Nikolai Ivanov, concurred: "We believed then that our work and sacrifices weren't in vain and that the time would come when people would set about turning the empty soil into a field of plenty."

Neither man, nor their colleagues, ever admitted in public to experiencing temptation during those cold, hungry weeks and months. A dishonest omission, a skeptical knower of human beings might conjecture, but one nevertheless borne out by the fact of their colleagues' deaths.

Ivanov and Lekhnovich never wavered from the conviction that they had made the right choice for the right reasons. In 1973 Ivanov, by then a gray-haired man of sixty-nine, gave an interview to the American journalist Georgie Anne Geyer, published in *International Wildlife* magazine. The pair met at the Institute. Geyer described how, as he answered her questions, Ivanov looked out his window "at the beautiful central square of Leningrad."

Ivanov described the evacuation of the field in Pavlovsk, the incendiary bombs that landed on the building's roof, the plague of vermin that attempted to eat the collection, and how he and his colleagues boarded up the shattered windows and protected the seeds from starving looters—desperate people who could not, presumably, understand why these scientists would choose death over survival. Ivanov described how he and his few remaining colleagues had planted two hundred varieties of seeds in the state farm under German fire, and how, while 20 percent of the collection gave no results after the siege ended, "the rest bloomed."

Geyer, perhaps not understanding the sensitivity of the question, asked Ivanov why the seed bank staff refused, even as they starved to death, to eat the nuts and seeds that were available to them. The bodies of the dead were scattered on the streets outside. It is undeniable that the seed stores could have prolonged the lives of the people of Leningrad through the siege's most difficult winter months.

"Why didn't we do it?" Ivanov replied. "Because our task was to save the collection. We knew that later, after the war was over, our country would need those seeds more than ever." Ivanov was, Geyer wrote in her article, "utterly convinced of the rightness of what they did."

Ivanov's response must be set in context: these are the recollections of people describing events and experiences for which they, as starving human beings, were not fully present. Deprivation had reduced their minds, both in mass and capacity for reasoning. Their focus had narrowed to the immediate needs of each moment; there was no strength left to interrogate a taboo. Their memories were subject to further stresses and pressures: the desire to honor those who did not survive, to retroactively justify decisions that carried fatal consequences, and to shape stories that enable the living to continue living.

Several years later Alexander Borisovich Borin, a journalist for *Literaturnaya Gazeta*, interviewed Ivanov and Lekhnovich.

"It is not enough just to glorify these people," he wrote. "We must also understand how they could die of hunger among food. What forces were at play? What did they think, what did they feel, what did they say?" Perhaps emboldened by Geyer's line of questioning (he references her piece in his story), Borin asked Ivanov why he and his colleagues had not eaten those seeds that had passed their expected germination date.

"My question seemed strange to Ivanov," Borin wrote.

"Why?" replied Ivanov. "There is a mandatory rule: to store samples not for five or six years, but at least ten to twenty years. The seeds age unevenly. Among the ten dead could be one alive."

Borin pressed the point, asking how many of these seeds then failed to germinate—"a cruel question," he admitted, but an important one. Ivanov replied that two metric tonnes of seeds failed to germinate.

"Two tonnes!" Borin replied. "Not eaten! Preserved in vain!"

"Why in vain?" Ivanov countered. "First of all, many of the samples that we thought were dead came up perfectly after the war. Besides, it's fortunate that these two tonnes of dead grain were in our hands as they made it possible for us to draw the most interesting conclusions. We learned that with the loss of grain germination, animals can no longer absorb the protein in the seed. . . . [This] helped change agricultural policy. So not in vain at all. No way."

For Ivanov, any scientific advance provided justification for the suffering of his colleagues. "What is truly being done for the sake of science cannot be fathomed," he added. "Never. We were well aware of this during the blockade. Otherwise, would we have had the strength to live?"

When Borin put the same question to Lekhnovich, he received an impatient response:

Imagine this scenario: Here you are, a writer, who has authored a book. You've put your all into it—your whole life. And suddenly, let's say, there is a severe frost, and you find yourself in a room without firewood to keep warm, only your manuscript. . . . Now can you begin to understand the psychology of the situation? You are freezing to death: Will you destroy this, the only copy of your book? Would you die to preserve this work? Yes, or no? Will you give in to temptation? What are you asking me, you and all the others? You're surprised? You're perplexed? Yes, it was difficult to walk at that time. It was unbearably difficult to get up every morning, move your hands and feet. . . . But to refrain from eating the collection? That wasn't difficult. No, not at all. Because it was impossible to eat your life's work, the life's work of your friends and colleagues. Do I really need to prove such an elementary, simple thing to you?

In the years that followed, Nikolai Ivanov worked tirelessly to revivify and honor his friend and teacher's work. He worked closely with Vavilov's widow, Yelena, to locate and collect Vavilov's unpublished manuscripts; the material Ivanov gathered facilitated the posthumous publication of three major books, including 1962's *Five Continents*, Vavilov's detailed account of his specimen-gathering expeditions around the world. Ivanov edited two editions of Vavilov's biography, and two articles that recentered his research in the arena of Soviet science.

Ivanov gave regular lectures at the seed bank, hoping to enthuse graduate students with his late teacher's theories, and ordered the instillation of a display board at the top of the main staircase at 44 Herzen Street, where he posted information and articles about Vavilov, including pages from his manuscripts, and letters and postcards from overseas expeditions sent to his colleagues describing the plant varieties he had seen, and offering recommendations and advice as an impetus to people's work and scientific publishing activity. At the Pavlovsk field station, where Kameraz and Voskresenskaya rescued potato samples under enemy fire, a mosaic portrait of Vavilov still decorates the staircase.

One mystery remained unsolved. What happened to the seeds taken by the Nazi commando?

VI

AFTER HE FLED LANNACH CASTLE, Heinz Brücher and his interpreter, Arnold Steinbrecher, found sanctuary with Brücher's parents near Heidelberg, Germany, about five hundred miles northwest of Lannach. For the next two years he kept a low profile. He evaded dispatch to Nuremberg to stand trial and drafted a substantial paper on barley enzymes, perhaps hoping to secure an invitation to the United States. As a former SS officer Brücher did not have the right to free movement in Allied-occupied Germany or Austria, but in 1947 he somehow managed to return to Lannach. It seemed he had indeed planted specimens in the hills and villages around Lannach. In a letter to his friend Theodor Herzog, he wrote, "This time I've been able to drive legally into Austria, to collect the remnants of the crops and scientific samples."

Not all the hidden seeds and plants were at Lannach when Brücher arrived. Venables did not seek to conceal the specimens he had smuggled back to Britain. The 1945 citation for the military medal he received mentioned "samples of wheat seeds, which agricultural experts believe may prove of inestimable value." Thereafter Venables set up an agricultural-seed merchant's in Chester; there is no record of whether the business was based on specimens drawn from Vavilov's stolen collection.[*]

Shortly after his trip to Lannach, Brücher chartered a fishing boat and sailed to Stockholm, where he remained in contact with Nazi-sympathizing Swedish scientists. When one Swedish geneticist raised the question of the whereabouts of the Lannach seeds, Brücher became, the scientist later recalled, "stiff"; another claimed Brücher showed him packets of seeds he claimed to have come from Lannach.

In November 1948 Brücher relinquished his German citizenship in exchange for a stateless passport. Then he flew to Buenos Aires with a third of a ton of extra luggage. In Argentina he began a career as a plant geneticist at the National University of Tucumán, where he became a prolific author of academic papers. Brücher remained an unreformed Nazi; when he visited South Africa during the apartheid era, he professed to an acquaintance he had seen evidence there to support his white supremacist theories. A staunch opponent of drugs and alcohol, Brücher boasted to the same colleague that he was working on a virus that could destroy the coca plant, from which cocaine derives.

---

[*] Venables suffered a stroke and died in May 1959 at the age of forty-four.

In 1958, before the story of Vavilov's murder had become known, the then director of the Plant Institute, Pjotr Zhukovski, visited Brücher in Tucumán. Ostensibly an innocuous meeting of two professional scientists, Zhukovski recorded that Brücher, "an experienced traveler, physically very fit . . . gave me a lot of help when I was in Jujuy Province." Before Zhukovski left, he invited Brücher to visit the seed bank in Leningrad, which, during the war, the German had been unable to reach. The former Nazi officer turned down the invitation but gave Zhukovski seven hundred varieties of wild potato to take back to the city. In the years that followed, the Institute staff would regularly receive packages of seeds in the mail from Brücher—a recompense, either given willingly or through coercion, for that which he stole as a younger man.

On December 17, 1991, police arrived at the Condor Huasi ranch in Mendoza, where Brücher—known to locals as Don Enrique—lived alone. Outside the white buildings stood an immense statue of a bird of prey, its five-foot wings outstretched in the style of the *Reichsadler* eagle. Inside, the police found German-language pornographic magazines, and piles of Nazi war memorabilia. Among them, facedown on the floor, lay Brücher, his hands and feet tied with electrical tape, which was also bound tightly over his mouth and nose. Brücher had died by asphyxiation.

The murder was dismissed as the result of a burglary; police arrested a Paraguayan man, then released him citing a lack of evidence. Years later Brücher's assistant Vincente Cabrera, who had collected seeds with Brücher since he was a teenager, told a reporter that Brücher had been working on a plan to eliminate the coca plant; he had been attempting to design a strain of the fungus *Fusarium oxysporum* to infect Bolivian coca plants with a virus that would not harm other vegetation. Ten days before Cabrera was due to travel to Bolivia to inject coca plants using a syringe, Brücher was murdered. "I'm sure that if I had been at the house that night, they would have killed me, too," Cabrera told a reporter.

With his death, Brücher took with him the information as to the whereabouts of the stolen seeds.

The Millennium Seed Bank, in Wakehurst, England, is a subterranean fire-, bomb-, and radiation-proof reinforced concrete bunker containing nearly 2.5 billion seeds. Like Vavilov's enterprise, the Millennium Seed Bank focuses on wild varieties of plant that might contain genetic traits and types of resilience that may yet prove to be valuable.

AFTERWORD

ON A WARM JUNE MORNING, the air humid with the threat of rain, I crunched into the car park at Wakehurst Place in the county of West Sussex, about an hour from my home on the south coast of England. There are several other botanical gardens like Wakehurst in the region: peaceful, idyllic grounds that orbit slightly dilapidated stately homes once owned by wealthy landowners. At some point, usually, the well of generational wealth ran dry, and these properties were bequeathed to the National Trust or a similar conservation charity, which subsequently maintains their grounds as tasteful heritage attractions for day visitors. Wakehurst is a little different. It has the mansion, the devotedly tended gardens, and the gift shop, but it is the only British public garden to feature a subterranean fire-, bomb-, and radiation-proof reinforced-concrete bunker containing nearly two and a half billion seeds, a vault known as the Millennium Seed Bank.

I had been instructed to meet Dr. Chris Cockel, a scientist who works at the seed bank, at the front desk at 10:30 a.m. I approached the ticket sales desk and awkwardly introduced myself to a man in a sleeveless green fleece.

"I'm afraid Chris isn't here yet," he said. "Would you mind waiting?" Then, nodding at the walkie-talkie on the desk in front of him, he apologetically added, "The scientists don't carry radios."

I considered buying a coffee at the nearby Seeds Café, but unsure if I'd be allowed to take a hot drink into a treasury of plant matter,

I decided instead to browse the sizable shop, which sold a range of kitchen-garden-related items: whiskey marmalade, brownie mug mixture, tiny succulents in brightly colored pots, wildflower meadow seed balls, and some Kew Gardens–branded jelly beans.

Wakehurst has been managed by Royal Botanical Gardens, Kew, since 1965, when the story of Nikolai Vavilov and his botanists had only recently begun to emerge. In 1976 staff from Kew, which is in South London, drove thirty miles to Wakehurst with the first delivery of seeds in the trunk of the car. These seeds were initially kept in Wakehurst's sixteenth-century mansion, then moved to a chapel on the grounds, in which the pews had been swapped for refrigerators. This meager collection formed the basis of the Millennium Seed Bank, which opened in 2000, at a cost of £20 million ($30 million).

I grew up a short walk from the gardens at Kew. My first part-time job as a teenager was in the café there, where I cleared tables alongside one of London's most prolific graffiti artists. On Saturday mornings I would pass Great Pagoda, an eighteenth-century folly designed by Sir William Chambers as a gift for Princess Augusta, the founder of the gardens. I'd walk between the rare plants and trees, then serve jacket potatoes and slices of pizza to tourists. "Kew Gardens sells jelly beans now," I thought.

After ten minutes or so Dr. Cockel had still not arrived, so the man in the fleece, who by now seemed a little agitated, approached and asked if I would like him to escort me to the seed bank.

"We'll probably run into Chris on the way," he said. "Ah, here he is now."

Dr. Cockel, wearing green chinos and a green shirt, shook my hand and led me along a path and around a corner to face the Millennium Seed Bank's glass expanse of frontage. He had joined the project, he explained, eight years ago and currently serves as the UK Conservation Projects coordinator. Most of the other seventeen hundred seed banks around the world hold agricultural species as insurance for an environmental calamity that, if it struck one of the world's major crops, could cause widespread food insecurity, even starvation. Wakehurst, by contrast, predominantly contains wild varieties of seed, gathered from 190 countries around the world.

The motivation behind the project is that which compelled Vavilov to tour the world: the belief that wild varieties of plant might contain genetic traits and types of resilience that may yet prove to be valuable. "By saving wild varieties we lock away traits that might be important in the future," Dr. Cockel told me. The Millennium Seed Bank holds "orthodox" species, seeds that are tolerant to drying and freezing, which excludes many tropical species (as well as seeds from oak and horse-chestnut trees) that break down during freezing.

At a time of unprecedented decline of natural habitats, Dr. Cockel and his colleagues, like Vavilov before them, feel a keen urgency to conserve the seeds of threatened plant species before they are lost forever.

"We are in a race against the loss of habitats," he explained. "It often feels like a race we are losing." Recently botanists in Malaysia wanted to make a collection of wild banana seeds. By the time they arrived to collect specimens the land had been cleared to make way for oil palm—*Elaeis guineensis*. Sometimes, however, the seed bank can restore collected specimens to habitats from which they have disappeared.

"Look at this!" Dr. Cockel told me, showing me a plant known by the inglorious name broad-leaved cudweed—*Filago pyramidata*. This is one of the rarest plants in the UK, a victim of fertilizers and herbicides, loss of fallow land, and development of highly productive crop varieties. Scientists at the Millennium Seed Bank recently grew 150 broad-leaved cudweed plants from which they eventually harvested three hundred thousand to be reintroduced to the English countryside, helping to restore and conserve biodiversity. The value of this work is not merely theoretical. It would cost an estimated £100 million ($125 million) to replenish the collection stored here, and as with Vavilov's seeds, some might prove irreplaceable.

"Vavilov and his teams were way ahead of their time," Dr. Cockel said, as we stood in the entrance hall to the seed bank, where the public can peer through glass windows into laboratories and watch scientists process seeds as they arrive from around the world. "They pioneered many of the processes we follow today."

In the first of these rooms the seeds undergo drying. A seed's lifespan doubles for every 1 percent its moisture content is reduced. Once dried out, seeds progress to the cleaning room, where botanists remove extraneous plant matter. Here a battalion of copper and stainless-steel sieves can be seen lined up on a shelf above a sink, as in a busy kitchen in a major city hotel. Finally, the seeds are x-rayed, the easiest nondestructive way to check if any samples are infested with insects or are half-eaten or not fully developed.

Not all the seed bank's samples are held in the vault, Dr. Cockel told me, beckoning for me to follow him to an outbuilding not accessible to the public.

Inside he led me into a sun-filled room, heavy with the sweet scent of warm plant matter, where plants of wild shapes and colors sit on wooden tables with wheels affixed to the tips of their legs. He gestured to a group of identical plants with fur-tipped leaves, grouped around a much larger specimen. Earlier, mindful of Ivanov's desperate mission to plant out specimens on the front line or else lose them forever, I had asked Dr. Cockel how long seeds could be kept before they lost their latent potential.

"This is *Leucospermum conocarpodendron*," he said.

More than two hundred years earlier, in March 1803, he explained, a Dutch explorer named Jan Teerlink had harvested this plant's seeds in Cape Town, South Africa. When he returned to his ship, the *Henriette*, Teerlink placed a handful of seeds inside his red leatherbound pouch, alongside silk gathered on an earlier trip to China.

On the return journey to Holland, the British warship *Lapwing* captured the *Henriette*, which was laden with tea from China. The navy took the requisitioned loot to London, where the red pouch was stored for two centuries. In 2005, a researcher at the National Archives happened upon the red pouch and discovered eight of the seeds inside. The samples were brought to Wakehurst, where, in 2006, three of the eight seeds germinated. They had retained their latent life for two centuries. All the plants on the table in front of us, Dr. Cockel explained, came from those three seeds.

We returned to the main seed bank building and, after I had signed in at the staff entrance, descended a narrow, winding metal staircase to the door that leads to the collection. The door sits as heavy and imposing as the entrance to a bank vault and includes a wheel that must be wound anticlockwise to disengage a series of thick bolts. We stepped through the door into a small antechamber, an air lock designed to maintain a steady temperature of 15°C (60°F) and 15 percent relative humidity. The vault door clunked behind us, then we walked into a gymnasium-size hall, which was mostly empty save for a woman who was seated at a table in the middle of the floor, using tweezers to pick up tiny seeds from petri dishes and place them into glass jars.

I noticed a clothing rack in a corner on which identical thick blue overcoats hung—protective wear for anyone entering one of the four doors that lead into the storage rooms themselves. Two of the four rooms house the Active Collection—seeds used for germination testing, and to be distributed to partner organizations—and the essential Base Collection. Coolers maintain a temperature of -20°C (-4°F) in all the four rooms. Sensors detect when a person enters, and an alarm will sound if the individual does not leave within ten minutes. A visitor who collapsed inside might soon freeze to death, even if wearing one of the overcoats. I was surprised at how modest the rooms were in size, especially when Dr. Cockel told me they are sufficient to store 75 percent of the world's flora.

"Currently we have just sixteen rooms," Dr. Cockel said. "That includes seventy-five percent of the UK's plant species."

I peered inside the windows to one of the storage chambers. Inside there were rows of shelves, just like those installed by Vavilov and his team in St. Petersburg. Instead of long metallic tins, however, these seeds are kept in the kind of glass kitchen jars inside which a cook might store rice or pasta. The seed bank uses Italian Bormioli jars, as they have particularly robust rubber seals. Inside each jar a packet of orange silica gel sits alongside the seeds; if the silica turns green, then moisture is present, indicating that the samples must be redried.

Dr. Cockel has not lost his youthful excitement for his work, the same glee felt by a collector of rare artifacts. The previous day a group of schoolchildren visited the seed bank. He gave them a rare marsh-land specimen to take with them and introduce to the school pond. "The teacher isn't a botanist," he said. "She teaches business—but she seemed committed."

Storing threatened plants is quiet, slow, often-thankless work. But for those who give their lives to the mission, it frequently becomes an obsession. I ask Dr. Cockel if he can understand the decision made by his forebears at the Plant Institute to refrain from eating the seeds they and *their* forebears had collected during years of expeditions.

"When you spend hours and days processing some of the world's rarest seeds, your heart and soul infuses the collection," he told me. "If I was starving to death, yes, I might be tempted to eat some of the samples. But I can equally understand the urge to abstain."

I HAVE NOT KNOWN HUNGER. Not properly. Like everyone else, I've been introduced: on a school trip when I ate my packed lunch soon after the coach departed, leaving only crumbs to see me through to dusk. Or on a long night drive when every fast-food stop had closed a few hours earlier and chewing gum could only stretch so far. Or at the cinema, when I arrived too late for popcorn. Hunger, but glancingly. Not close enough to shake its bony hand, then feel those fingers begin to rub the flesh away.

I know, vaguely, that hunger squats, ever patient, in the empty space on the supermarket shelves behind the tins of soup and boxes of cereal. I know that it waits for the systems to snap, the logistics to fail, when it finally emerges, to move in intimately and, given time, fatally.

How long might you or I survive if hunger arrived today? In my house there are currently four mouths to feed—five if you count the dog (and for how long could we count the dog?). The cupboard's inventory would not put up much of a barricade: four packets of dried penne; two tins of tuna; a little glass jar of mixed herbs; half a pack of butter. The vegetable patch—something fun to do with the kids!—might give us a week or two extra with its meager harvest of buried onions, slug-nibbled lettuces; a beetroot left to balloon and toughen for too long in the ground. September. October. November. December. By then, I'd know hunger well; perhaps it would already have been the last thing I knew.

In St. Petersburg, a person today might not know hunger personally, but in many cases hunger is known to the family. There are the private stories, preserved and passed down like instructive heirlooms, of the grandfather who relinquished his ration to save a neighbor's child, of the great-aunt who collected drinking water from a crack in the ice on the Neva, of the sibling who perished on the ice road. Each successive generation has been born into the shade of the enduring and long shadow the siege continues to cast across the city.

There are the public stories, contained in monuments, the smoothed stone artifacts of Soviet public remembrance designed, as most war memorials are, to present suffering as sacrifice. Everyone in the city has at one time walked the birch-lined Avenue of the Unvanquished, which leads toward Piskaryovskoye Cemetery, the site of the 186 mass graves that hold half a million civilians and soldiers who died during the siege, among them a few botanists.

There, etched in granite, the words of the poet Olga Bergholz, adopted as the city's motto:

KNOW THIS, YOU WHO REGARD THESE STONES: NO ONE IS FORGOTTEN, AND NOTHING IS FORGOTTEN.

It's a worthy aspiration motivated by the same impulses that move the historian, the diary writer, and the genealogist. And yet, as any of these people could tell you, it is an impossible goal. From the great churning cascades of time, few stories are ever fully grasped and saved, and never completely. Most is lost. That which survives does so in fragments, easily corrupted by the slow erosion on retelling, even by the protagonists.

I first learned about the activities of the seed-bank staff when I read an article published in the *Guardian* newspaper about the palace gardens in Pavlovsk. By then this area south of the city, where Kameraz staged his potato rescue attempt, housed more than five thousand varieties of fruits and berries grown from seeds collected by Vavilov and his teams.

This living repository of trees and bushes, the article stated, was now embroiled in a dispute as a federal Russian housing agency tried to confiscate part of the Pavlovsk research station to clear the land for upscale dachas. The previous year the Russian Ministry of Economic Development handed over one-fifth of the station's fields to the Federal Fund of Residential Real Estate Development as housing land. The Plant Institute appealed the decision. Fyodor Mikhovich, the Pavlovsk station's director, who had worked there for thirty-two years, claimed an official told him, "Go to sleep. Just go to sleep. We are taking the land."

The article mentioned how, during the siege, scientists at Vavilov's Institute had protected its collections, with some succumbing to starvation rather than consuming the collection's rice and other crops. And while today the Institute and its experimental stations have become "undeniably dilapidated," staff continued to maintain the collection of old varieties.

The following month the French newspaper *Le Monde* published a follow-up: "Russia backs away from plans to break up the unique Pavlovsk seed bank." But while the gardens would not be used for luxury flats, the piece included a disheartening aside about the current state of the station: The wages are pitiful, averaging about $270 per month; equipment is scarce, and the administrative buildings are almost a ruin. City people joke about the botanists and geneticists at the Institute being prepared "to fork out to be able to work there."

Public complaints about the state of funding at the seed bank were not new, either. In a 2008 article published in the Russian edition of *Newsweek*, Viktor Dragavtsev, a former director of the Institute, complained about the lack of safety equipment in the building. "The Petersburg collection isn't free from risk," he told the journalist. "There's no fire alarm system, so any ordinary fire in the building could destroy the collection in three hours."

I wrote to the seed bank in St. Petersburg to ask whether they might permit me access to the Institute's archives. Apart from Senin's novella-length 1979 work, *List'ya vyrastut vnov'* (Leaves will grow again), there is no book, neither in English or in Russian, that has attempted to tell the story of the scientists during the siege in proper context. So began a lengthy exchange of emails, Zoom calls, and official letters to negotiate a visit to St. Petersburg and the Institute's archives.

I planned to visit the seed bank in the summer of 2022 and employed the services of a genealogist named Kirill, to whom I paid £200 ($225) to make some initial inquiries to find descendants of staff who worked at the seed bank during the siege. Together we compiled a list of twenty-five staff members for the initial investigation. Kirill told me he needed a month, then we should plan to meet in the city "for further talks, interviews, and research."

On February 24, 2022, I was en route to London to attend a memorial service for a friend when I heard a reporter for BBC Radio 4 announce the Russian invasion of Ukraine. Lines of static traffic wended their way out of Kyiv, he explained, as citizens sought to flee the city and country. Russia's assault began with a series of missile attacks and the use of long-range artillery. It quickly spread across central and eastern Ukraine as Russian forces attacked the country from three sides. Russian forces were, by the following morning, twelve miles from the Ukrainian capital. That same day, Russian forces began to stage a siege of the Ukrainian city Mariupol, situated in Donetsk Oblast. The Red Cross would later describe the situation in Mariupol as "apocalyptic"; Ukrainian authorities accused Russia of engineering a major humanitarian crisis in the city.

That evening I checked the government website for advice to British citizens hoping to travel to Russia. "The Foreign Office advises against all travel to the whole of Russia due to the lack of available flight options to return to the UK, and the increased volatility in the

Russian economy," it read. "If your presence in Russia is not essential, we strongly advise that you consider leaving by remaining commercial routes." Kirill stopped answering my emails.

Olga Valerievna Yusim, the leading specialist at the Plant Institute, with whom I had first started communicating, continued to reply to my messages. She suggested we move our conversation to the secure messaging service Telegram. Mindful that our communications might be seen by others, we did not directly discuss the "military operation" in Ukraine, as Putin's government euphemistically referred to the war, "Situation is changing every day," she wrote. "We have a new pestilence worse than COVID now. I don't know when and how it will end up."

Olga, who has shoulder-length brown hair and a faraway smile, and who talks slowly, as if weighing the implication of each word, began to scan and photograph dozens of books, articles, staff papers, photographs, and other documents from the Institute's library and archive relating to the period to assist my research. She took photographs of commemorative plaques on the walls; she sent pieces written by the siege botanists, both published and unpublished.

The seed bank's musty corridors, seed tins, and broken-down lifts remained out of reach. But the documents—as documents often do when met carefully, like old, undisturbed rooms—had a transposing effect. By combining these with the dozens of diaries and first-person accounts, as well as a red-backed copy of Popovsky's biography (a copy of which my father, a book collector, tracked down in America after I had searched for weeks in vain), a picture of life in the seed bank in the city emerged from the mists that hang, infuriatingly, just beyond living memory. An incomplete picture, yes, but a useful and compelling one all the same.

Still, here at the end of the journey, frustrations remain, as every historian and writer of nonfiction must privately admit. We leap between stepping stones of known fact, around which lap the shallows of speculation. To be precise: straightforward questions remain unanswered by the restrained testimony of the few survivors who recorded their experience. In the Hermitage, across from the Plant Institute, the frozen corpses of the dozens of employees who died in the building's cellars were stored in a museum room until they could be removed. Did the bodies of Alex Shchukin, Dmitri Ivanov, and all the others remain similarly in situ. If so, for how long? Who removed them? Who watched on?

Was Ivanov, who argued forcefully against consuming any specimens from the collection, ever haunted by the deaths of those who upheld his executive decision—made in opposition to the instruction of his superiors, who urged the botanists to "spare nothing"? Were Institute staff eligible for the more generous state-controlled rations made available to some other essential workers?

Behind these practicalities, emotional questions loom, too. When the snow began to thaw and revealed the bodies that had been consumed by ice in the terrible winter of 1941, how did the surviving staff members mourn their dead? Was their clarity and force of purpose sufficient to push these thoughts from their minds? What animosity existed between the remaining staff and Vavilov's replacement, the Lysenko-supporting (according to Mark Popovsky, at least) Johan Eichfeld, former director of Vavilov's Polar field station, who remained Vavilov's direct successor until 1951? And how did Ivanov feel about overseeing the Institute while the whereabouts of his former superior, old friend, and longtime supporter, Vavilov, were unknown?

When, in the autumn of 1944, a group of Institute staff returned to Leningrad from the Ural Mountains, what was the emotional frequency of their initial encounter with the other survivors? Was there forevermore a rift between those who escaped and those who remained? If there existed, as there surely did, a special bond between those who had endured the siege inside the city, how did that intimacy manifest in professional life, and for how long? What was felt and said and remembered with the final confirmation of Vavilov's death? What else died that day?

All these memories are lost to the raging currents of time. Even with Olga's help and descriptions, scans of the original orders and directives sent to the Institute during the siege, the testimonies of former employees, the recollections of Vavilov's friends, and the dozens of photographs of the rooms in which the collection was stowed (which today look much as they did in 1941), I often felt estranged from the world I was so desperately attempting to inhabit. Rereading this manuscript repeatedly (unlike poetry, prose is never finished; you merely reach the point where you think, "That will do, and it must"), I began to see only the gnawing absences of fact about what happened, and beneath that, the absences of clear justification for what happened.

To strengthen my connection from another part of the world to these people, who lived and died decades before I was born and in circumstances I can only imagine, I decided I would plant some seeds. In the front garden outside my house, I laid railway ties down, screwed them together to form three large rectangles, and filled them with a mixture of topsoil and compost. I planted seeds in tiny cardboard cups designed to encourage germination and placed them in a plastic "growhouse" that I ordered from Amazon—a capitalist extravagance that only distanced me further, I later felt, from the Communist-era botanists whose example I wanted to follow.

I listened to Monty Don, the soft-voiced, big-jumpered mensch of the BBC's weekly *Gardeners' World* television program and diligently followed his advice in the hope of replicating his outlandishly bountiful harvests. After the seeds germinated, I planted them out into the trio of beds. In one I placed the marrows, zucchini, artichokes, and onions. In another a raspberry bush, and neat rows of strawberries. In the third, I planted seed potatoes from a packet that claimed they would be ready for Christmas Day. I watered the patches with indulgent regularity and marveled as some of the plants overtook others and began to produce fruits and vegetables. These seeds had been tweaked and refined by selective growing across many decades, building, if not on the varieties collected by Vavilov and his team, then certainly on the principles they had honed through their pioneering work. I felt a keener connection to the group, not through their writings, which so often felt discreet and controlled, but through this relation to the soil, to the elemental act of growing the foodstuffs on which life itself depends.

Still, these efforts could not transport me to the white-hot center of the story, to the unspeakable question that underpins this forbidden garden: Was the sacrifice worth it? Did these men and women, in opting to sacrifice the lives of real people for the benefit of the imagined many, make the moral choice? I understand why, as Ivanov put it to the journalist, people could not destroy their irreplaceable life's work, why they might choose to perish so that this legacy might live on. But to do so at the cost of people's lives other than their own? To this I have no answer.

IN THE SUMMER OF 2023, shortly after I visited the Millennium Seed Bank, I wrote to Elena Khlyostkina, the current director of the Institute. I wanted to know what the story of the botanist Nikolai Vavilov's suffering and sacrifice meant to her, as the person who now occupies the same seat once held by him. I wanted to ask whether she believed Ivanov and his colleagues made the right decision.

I sent the email, then waited. It was an unusually disruptive moment in the Russian-Ukrainian conflict. On June 23, 2023, Yevgeny Prigozhin, leader of the Wagner mercenary force, had staged an attempted coup. The group took control of Rostov-on-Don and then continued toward the Kremlin, apparently hoping to seize power. Then, after Vladimir Putin addressed the nation to claim that Prigozhin had stabbed Russia in the back, the mercenary appeared to reconsider. He called off the advance and claimed refuge in Belarus. (Two months later, after ostensibly making peace with Putin, Prigozhin died in a plane crash, which Russian authorities claim was an accident.)

Khlyostkina did not respond for reasons that are unclear but easily guessed: to correspond with a Western author and journalist at a time of national crisis could surely invite needless difficulties upon her and her institution.

Sensing my disappointment, Olga sent me a brief television documentary about the seed bank in which Khlyostkina features. At various points in the film, we see the director speak to the camera. In one section she describes how, behind her chair in her office at the Plant Institute, there hangs a huge portrait of Nikolai Vavilov.

"Becoming director of VIR, I obviously have a huge sense of responsibility," she said. "Even the portrait on the wall behind my desk is a help because I know that no matter how difficult things get, and without wanting to make invidious comparisons, it's nothing compared to the physical and emotional suffering [of Vavilov and colleagues]. It makes our difficulties seem trivial."

ACKNOWLEDGMENTS

SPECIAL THANKS TO ANDREW SPENCER, Marina Zaretskaya, and Anna Nyburg for their extensive and careful support in translating various Russian- and German-language documents into English.

Special thanks to Dr. Uwe Hossfeld from Friedrich-Schiller-Universität Jena for his generous sharing of research materials and personal correspondence with Heinz Brücher's translator, Arnold Steinbrecher.

Special thanks to Olga Valerievna Yusim for her untiring help, advice, patience, and provision of access to original notebooks, diaries, and other documents held at the Plant Institute, and to Professor Igor Loskutov for sharing information and knowledge.

Thank you to Liz McGow, archivist at the Linnean Society of London, for her assistance in locating documents in the collection relating to Nikolai Vavilov and the VIR. Thank you to Tara Craig, head of public services at the Rare Book & Manuscript Library, Columbia University, for her assistance in accessing Mark Popovsky's papers.

Thank you to the author and journalist Peter Pringle for his generous sharing of information and contacts. Thank you to my father for helping locate an original copy of *The Vavilov Affair*, Popovsky's rare and long out-of-print 1984 book, and to Rosemary Marshall for sending it on when it was delivered to her address in Sarnia, Ontario, by accident.

Thanks to Sebastian Kettley and Dr. Chris Cockel from the Royal Botanic Gardens, Kew.

Thank you to the genealogist Kirill Chashchin.

Thank you to my editors Juliet Brooke, Charlotte Humphery, Emily

Polson, and Colin Harrison, and to my proofreaders Daniel Cuddy, Nico Parfitt, and Helen Parham.

The *Forbidden Garden* was written with the support of grants awarded by the Sloan Foundation, in the United States, and the Society of Authors, in the United Kingdom.

STAFF ROLL CALL

(July 1941–45)

THE FOLLOWING INDIVIDUALS ARE CONFIRMED to have worked at the Plant Institute during the siege of Leningrad. This information is drawn from various sources, including original papers held in the Institute's archives, written orders issued by the Leningrad authorities, diaries, memoirs, and a variety of secondary sources. Some staff members were made redundant during this time, others were evacuated from the city in February 1942, while some died either in the seed bank, at their homes, or while fighting at the front. Where a date and cause of death is known, it has been recorded here, in most cases for the first time in a published book.

KEY TO SOURCES

Evac.—evacuated from the city.

FF—firefighters.

Gol.—named by G. Golubev in his 1987 book, *The Great Sower.*

Iva.—named by N. R. Ivanov in *V Osazhdennom Leningrade* and/or an unpublished memoir.

LADF—member of the Local Air Defense Force.

Lek.—named by Vadim Lekhnovich in *V Osazhdennom Leningrade.*

O25—named in Order 25, signing off for the evacuation of research staff, February 14, 1942.

Pav.—named by Dmitri Pavlov in his 1965 book, *Leningrad 1941.*

Pop.—named by Mark Popovsky in his 1984 book, *The Vavilov Affair.*

Sen.—named by Viktor Senin in his 1979 book, *List'ya vyrastut vnov'.*

VIR—commemorated on a board at Institute as one of nineteen staff members who "died in office."

NAME	ROLE	DATE & CAUSE OF DEATH	SOURCE	PHOTOGRAPH
Andreyeva, Anna Pavlovna	*Caretaker*		FF; Iva.; Sen.	
Antropova, V. F.	*Crop scientist, expert in rye, Institute employee from 1925*		Iva.; Pav.	
Babichev, I. A.	*Senior research officer (in charge of material and equipment for the biochemistry lab and civil defence at Building 42, Herzen Street)*		O25	
Baikov, Andrey Ivanovich	*Institute's driver and mechanic*	c. 1941–42, starvation	VIR	
Bazhanov, I. S.	*Cashier*	c. 1941–42	VIR	

NAME	ROLE	DATE & CAUSE OF DEATH	SOURCE	PHOTOGRAPH
Belyaeva, Maria Sergeyevna.	*Institute building manager*		FF; Lek; Iva; Sen.	
Biryukova, M. E.	*Caretaker*		Iva.	
Bogdanov, P. A.	*Unknown*		Iva.	
Bogushevsky, Pyotr N.	*Fruit expert*	c. 1942 starvation	Evac.; Iva.; Sen.	
Brezhnev, D. B.	*Head of dept. of vegetable crops, 1937–41; sent to the front in 1941; director of Plant Institute, 1965–78*			
Chernyanskaya, Klavdiya Mikhailovna	*Senior lab technician, maintenance of the root-vegetable collection, fruit cultivars, and root-vegetable fodders*		Lek.; Pav.; Sen.; O25	
Dmitricheva, Maria Parfinievna	*Librarian*	Nov. 11, 1941, starvation	VIR	
Dmitrieva, E. I.	*Institute accountant*	Mar. 15, 1942, starvation	O25; VIR	

NAME	ROLE	DATE & CAUSE OF DEATH	SOURCE	PHOTOGRAPH
Egiz, Prof. Samuil Abramovich	*Head of tobacco and tea dept.*	Early 1942, starvation	Iva.	
Eichfeld, Johan G.	*Plant Institute director, 1940–51*		Evac.; Iva.; Sen.	
Gleiber, E.	*Archivist*	c. 1941–42, starvation	Gol.	
Golenishchev, Grigori Ilyich	*Caretaker*	Nov. 1941, shelling	FF; Iva.; VIR	
Golubyova, L. A.	*Second deputy chief accountant*		O25	
Gufeld, A. I.	*First deputy chief accountant, in charge of completing the annual accounts for the center, left for Krasnoufimsk on submission of the accounts*		O25	
Gusev, Pavel Pavlovich	*Institute adviser, Vavilov's former secretary*	Nov. 11, 1941, starvation		
Heintz, Georgi Viktorovich	*Head librarian*	Jan. 16, 1942, starvation	Iva.; LADF; O25; Sen.; VIR	

NAME	ROLE	DATE & CAUSE OF DEATH	SOURCE	PHOTOGRAPH
Ikkert, I. A.	*Laborer, in charge of maintenance of property at the civil defense station in Ovtsino*		O25	
Ivanov, A. I.	*Caretaker, laborer*		Iva.	
Ivanov, Dmitri Sergeyevich	*Head of cereal crop and rice dept.*	Jan. 9, 1942, starvation	Gol.; Iva.; LADF; Sen.; VIR	
Ivanov, Dr. Nikolai Rodionovich	*Head of biochemistry laboratory from 1922*		Gol.; Iva.; Pav.; Pop.; Sen.	
Ivanova, P. A.	*Unknown*		FF; Iva.	
Kameraz, Abram Yakovlevich	*Senior researcher, potato specialist*	Jan. 15, 1991	Lek.	
Kapustina, Ekaterina	*Wheat dept.*			
Katkova, Nadya	*Laboratory assistant at the department of cereals, Komsomol secretary*		Iva.; Sen.	

STAFF ROLL CALL   315

NAME	ROLE	DATE & CAUSE OF DEATH	SOURCE	PHOTOGRAPH
Kipl, Yelena. S.	*Junior research officer, maintenance of cereal crops: barley, oats, wheat*		O25	
Kordon, Rudolf Yanovich	*Senior research officer, apple expert*	1961	Iva.; Pav.; Sen.; O25	
Korzun, A.	*Unknown*	Starvation		
Kovalenko, A. P.	*Lab technician*		Lek; Sen.	
Kovalevsky, Dr. Georgi Vladimirovich	*Researcher*	Jan. 1942	Sen.; VIR	
Kreier, Dr. Georgi Karlovich	*Specialist, medicinal herbs*	Jan. 12, 1942, starvation	Iva.; Sen.; VIR	
Kurgatnikov, M. M.	*Biochemist*	Died at front	Iva.	
Kuznetsova, Evgenia Sergeyevna	*Head of the section of forage grasses*	Nov. 19, 1983		

NAME	ROLE	DATE & CAUSE OF DEATH	SOURCE	PHOTOGRAPH
Lavrova, M. N.	*Researcher*	Mar. 22, 1942, starvation	VIR	
Lebedeva, Galya, A.	*Laboratory technician*		Lek.; Sen.	
Lekhnovich, Vadim Stepanovich	*Curator of tuber collection*	1989	Gol.; Iva.; Pav.; Sen.	
Leontievsky, Dr. Nikolai Petrovich	*Senior researcher of agrometeorology*	Jan. 5, 1942, starvation	LADF; VIR	
Likhvonen, Nikolai Nikolaevich	*Procurement agent for Institute*	Early 1942	LADF; VIR	
Luss, A. I.	*Fruit specialist*	Died at front	Iva.	
Malgina, Anisiya Ivanovna	*Head archivist*	Early 1942, starvation	FF; VIR	
Mikhyever, P. N.			Pav.	

NAME	ROLE	DATE & CAUSE OF DEATH	SOURCE	PHOTOGRAPH
Miloslavskaya, Natalia Efimovna	*Senior lab technician, maintenance of the fodder collections*		O25	
Moliboga, Alexander Yakovlevich	*Senior researcher at the dept. of agrometeorology*	Jan. 1942, died in fire caused by shelling	Iva.; Sen.; VIR	
Mordvinkina, A. I.			Pav.	
Panteleyeva, Klavdiya Andreyevna	*Senior research officer, asters expert*		Iva.; Pav.; Sen.; O25	
Pavlova, Evdokia Fyodorovna	*Research officer, in charge of maintenance of material and equipment in the dept. of agrometeorology*		O25	
Pavlovich, Sonya	*Unknown*		Iva.	
Petrova, Praskovya Nikolaevna	*Research officer, in charge of property for seed monitoring, the laboratory, and sanitation*		Iva.; Pav.; Sen.; O25	

NAME	ROLE	DATE & CAUSE OF DEATH	SOURCE	PHOTOGRAPH
Reuter, Georgi Nikolaevich	Chief administrator and head of personnel		Iva.; Sen.; O25	
Rodina, Lidia Mikhailovna	Associate researcher, keeper of the oat collection	Jan. 1942, starvation	Gol.; Sen.; VIR	
Romanova, M. F.	Unknown		FF; Iva.	
Rubtsov, Grigori Aleksandrovich	Senior researcher, fruit crop dept.	Apr. 14, 1942, during evacuation	Evac.; FF; Sen.; VIR	
Semyonov, V. M.	Head of stores, in charge of the research section's stores and house storekeeper		O25	
Shchavinskaya, Serafina Arsenievna	Associate researcher	Apr. 14, 1942, starvation	VIR	
Shcheglov, Mikhail Andreyevich	Trained agronomist, worked as laborer at Institute during blockade for ration card	c. 1941–42, starvation	Sen.	

NAME	ROLE	DATE & CAUSE OF DEATH	SOURCE	PHOTOGRAPH
Shchukin, Aleksandr Gavrilovich	*Associate researcher of the collection of industrial and forage crops*	Dec. 25, 1941, starvation	Gol.; Iva.; Sen.; VIR	
Shebalina, Maria	*Researcher, dept. of fodder crops*		Pop.	
Tarasenko, G. G.	*Fruit specialist*	Died at front	FF; Iva.	
Tumanov, I. I.	*Unknown*		FF; Iva.	
Varavanova, I. I.	*Accountant*		025	
Virs, Yan Yanovich	*Deputy director*		Evac.; Iva.; Lek.; Sen.	
Volkova, M. F.			FF; Iva.	
Voskresenskaya, Olga Alexandrovna.	*Researcher, potato expert*	Mar. 3, 1949, complications related to shelling	Gol.; Iva.; Lek.; Pav.; Sen.	
Voyko, Elizaveta Nikolaevna	*Deputy head librarian*	Nov. 23, 1941, starvation	FF; Sen.; O25; VIR	
Wulf, Yevgeni Vladimirovich.	*Head of the herbarium, and geography, taxonomy, and ecology depts.*	Dec. 21, 1941, shelling	Gol.; Iva.; Sen.	
Yakubintser, M. M.	*Wheat expert*	Survived evacuation in Feb. 1942	Evac.; Pav.	
Yurtsev, I. Ya.			FF; Iva.	
Zaytseva, Galya	*Lab technician*		Sen.	
Zhernakov	*Technician*		Sen.	

NOTES

PROLOGUE

3 **nodded to a standstill:** Between 1914 and 1924, St. Petersburg was known as Petrograd, as during the First World War the original name was deemed too Germanic.

4 **"a treasury of *all known* crops and plants":** Letter to G. S. Zaitsev, 1920, as quoted in Loskutov, *Vavilov and His Institute*, 18.

5 **starving to death:** Kolchinsky, "Nikolai Vavilov in the Years," 334.

6 **"wounded look":** As quoted in Volkov, *St. Petersburg*, 224.

6 **"the ravens are in flight":** Akhmatova, *Selected Poems*, 49.

6 **an internal courtyard:** Fedotova and Goncharov, *Byuro Po Prikladnoy Botanike*, 7. The building that housed the original Bureau of Applied Botany is now the River Palace Hotel.

7 **and a foyer:** "About Us," www.riverpalacehotel.ru, retrieved June 2023.

7 **meaningful collection:** Loskutov, *Vavilov and His Institute*, 2–7.

7 **thoughtless invader:** Popovsky, *Vavilov Affair*, 32.

7 **"almost complete destruction":** Ibid.

7 **"a million problems":** Ibid.

I. AN EXPLORER VANISHES

11 **fifty-two-year-old leader:** Nikolai Vavilov, DOB November 25, 1887.

11 **urged his friends:** Golubev, *Great Sower*, 137.

12 **42 and 44 Herzen Street:** Now Bolshaya Morskaya, following the collapse of the Soviet Union and the reversion of street and place names to their pre-Revolution forms.

12 **"collection in the world":** "Plant Breeding in Russia," *Times*, December 4, 1933.

12 **"world collection of plants":** e.g. Roll-Hansen, "Wishful Science: The Persistence of T. D. Lysenko's Agrobiology in the Politics of Science", 166–88.

13 **to breed supercrops:** *Research Report of the All-Union Plant Institute* (VIR, 1945), 5.

13 **"sunshine and courage":** Ibid.

13 **"more life-loving, life-giving":** Letter from H. Muller to M. Popovsky, dated July 16, 1966, personal archive of Yuri Vavilov, as quoted in Loskutov, *Vavilov and His Institute*, 54.

13 **grand staircase:** Popovsky, *Vavilov Affair*, 119.

14 **"The brakes are getting worn down":** Ibid.

14 **Moscow, Leningrad, and Pushkin:** Hawks, *N. I. Vavilov Centenary Symposium*, 6.

14 **promising morning:** Popovsky, *Vavilov Affair*, 127

14 **"bright, intelligent eyes":** Lidia Petrovna, quoted in ibid., 17.

14 **"kings and queens":** Semyon Reznik and Yuri Vavilov, "The Russian Scientist Nikolai Vavilov," in Vavilov, *Five Continents*, xxi.

14 **shoots of talent:** A natural mentor, Vavilov invested not only in young Soviet men and women but also their Western counterparts, including the American student John Niederhauser, who later won the World Food Prize, and the British student J. G. Hawkes, whom Vavilov took on expeditions into the countryside around Leningrad and invited to dine and attend the opera with his family.

15 **"the South Pole":** *Nikolai Ivanovich Vavilov. Scientific Legacy. From the Epistolary Heritage, 1929–1940*, vol. 10, of *Scientific Legacy of Nikolai Ivanovich Vavilov* (Moscow: Soviet Academy of Sciences and Nauka Publishing House), 120, 133, 135.

15 **"only second to bread":** Senin, *List'ya vyrastut vnov'*.

15 **then in potatoes:** "Vadim Stepanovich Lekhnovich," staff files, VIR.

15 **swashbuckling saint:** See, for example, Nabhan essay in Del Curto, *Seeds of the Earth*, 54.

15 **A polyglot:** Vavilov spoke fluent English, French, and German, and passable Spanish and Italian.

15 **eighteen-hour days:** Popovsky, *Vavilov Affair*, 19.

15 **"seldom have been matched":** As quoted in Hawks, *N. I. Vavilov Centenary Symposium*, 5.

15 **"One must hurry":** Ibid.

16 **three vehicles:** Another source described the cars as Soviet-made Ford sedans, which, if accurate, would likely have been the GAZ M1, a passenger car based on the Ford model and produced in Russia from 1936.

16 **make room:** Bakhteev, *Nikolay Ivanovich Vavilov*, 215–16.

18 **undergraduate thesis:** Baranski, "Nikolai Ivanovic Vavilov," Embryo Project Encyclopedia (EPE).

18 **eight years earlier:** "The History of Plant Science and Microbial Science at the John Innes Centre," www.jic.ac.uk.

19 **"evolution are hidden":** N. I. Vavilov, "The Process of Evolution in Cultivated Plants," *Proceedings of the 6th International Congress of Genetics* 1 (1932): 331–42.

19 **study cereals:** Hawks, *N. I. Vavilov Centenary Symposium*, 4.

19 **samples in Mongolia:** Del Curto, *Seeds of the Earth*, 4.

19 **Puerto Rico:** Vavilov, *Five Continents*, 17ff.

19 **regions and climes:** Bakhteev, *Nikolay Ivanovich Vavilov*, 215–16.

19 **sent home:** Golubev, *Great Sower*, 115.

19 **"inquiry and optimism":** As quoted in Hawks, *N. I. Vavilov Centenary Symposium*, 5.

19 **varieties he had seen:** Nikolai Rodionovich Ivanov, staff file, VIR.

20 **"on the globe":** Ibid.

20 **famine and starvation:** Popovsky, *Vavilov Affair*, 88.

20 **he called friends:** See the introduction in Vavilov, *Five Continents*, viii.

21 **within three years:** Nabhan, *Where Our Food Comes From*, 180.

21 **"from a hen's egg":** Popovsky, *The Vavilov Affair*, 52.

21 **"in the Institute's cupboards":** Soyfer, *Lysenko and the Tragedy of Soviet Science*, 69.

22 **expensive luxury tourist trips:** Kolchinsky, "Nikolai Vavilov in the Years," 4.

22 **"Not a thing":** Abram Yakovlev, "On the Theories of True Geneticists," *Socialist Reconstruction of Agriculture* 12:55–56.

22 **"Napoleonic in character":** S. Harland and C. Darlington, "Prof. N. I. Vavilov, For. Mem.R.S.," *Nature* 156 (1945): 621–22.

22 **use coded language:** Nabhan, *Where Our Food Comes From*, 181.

22 **Party affiliation:** Kolchinsky, "Nikolai Vavilov in the Years," 337.

22 **reputation by association:** Prezent, "Pseudoscientific Theories."

22 **"justify the expenses":** As quoted in Medvedev, *Vzlet i Padenie Lysenko*, 348.

23 **the biologist Cyril Darlington:** Ings, *Stalin and the Scientists*, 298.

23 **call home:** Savitsky, "My Remembrance."

23 **"impossible for me":** Vavilov, *Ocherki y Vospominaniya*, 367.

24 **on the pavement:** F. K. Bakhteyev, "The Last Decade," speech presented at the Moscow Society of Natural Researchers, November 24, 1965.

24 **"sudden recall to Moscow":** Note in possession of V. S. Lekhnovich, VIR.

24 **"Is it worth your while going?":** Bakhteyev, "Last Decade."

25 **Kobstev and Koslov:** Kobstev and Koslov are the two names recorded on the warrant, issued on August 7, 1940, the day *after* Vavilov's arrest in Ukraine.

II. UNWANTED TREASURE

28 **keep it to himself:** Beevor, *Stalingrad*, 4.

29 **until the war ended:** Up until the night of the attack, the Soviet Union's trade commissariat had in fact continued to send food and supplies to Germany, as part of the nonaggression pact between the two countries, thereby strengthening the soon-to-be attacker, and depleting Leningrad's reserves.

29 **Lieutenant Schmidt Bridge:** Nikolai Rodionovich Ivanov, staff file, VIR.

29 **thirty-nine-year-old:** Nikolai Rodionovich Ivanov, DOB May 29, 1902.

30 **publishing house had closed:** Loskutov, *Vavilov and His Institute*, 101.

31 **he would enlist:** Senin, *List'ya vyrastut vnov'*.

32 **stopped by a window:** Salisbury, *900 Days*, 130.

32 **Old Testament prophet:** See, for example, the sketch of Orbeli by Georgi Vereisky, 1942, State Hermitage Museum, St. Petersburg.

32 **eleven o'clock:** Salisbury, *900 Days*, 86.

32 **six calls:** Ibid., 130.

32 **willing to act:** In fact, like Orbeli, Colonel Ye. S. Lagutkin of the Leningrad Antiaircraft Command had been unable to reach anyone in Moscow. To direct antiaircraft units to their posts without risking accusations of acting without orders, he ordered a "practice" drill.

33 **packaged for safekeeping:** Orbeli's motives for these preparations remain unclear. Other than the joiners, only the head of the museum's special department, Alexander Tarasov, was informed of the plans.

33 **million coins and medals:** Varshavsky and Rest, *Saved for Humanity*, 479.

34 **"house after a funeral":** Quoted in Reid, *Leningrad*, 60. Original source, now deleted: www.hermitagemuseum.org.

34 **received a warning:** Popovsky, *Vavilov Affair*, 163.

35 **"trifling matters":** Ibid.

35 **interests of science:** Testimony of Maria Shebalina, quoted ibid.

35 **"empty, tired eyes":** Ibid., 164.

35 **"no misunderstanding":** Ibid.

36 **promoted scientists:** The Soviet-era journalist Mark Popovsky described Eichfeld

as "an able agronomist with a good knowledge of biology, but a faithful follower of Lysenko."

36 **promptly died:** Professor Sinskaya in Popovsky, *Vavilov Affair*, 159.

36 **foghorns of ships:** Senin, *List'ya vyrastut vnov'*.

37 **"stomach is stretchable":** Kay, *Exploitation, Resettlement, Mass Murder*, 167.

39 **resting their paper:** Ibid., 219.

39 **called upon if required:** Senin, *List'ya vyrastut vnov'*.

39 **drape blankets:** See, for example, "siege room" exhibit at the Museum of Bread in St. Petersburg.

39 **enemy propaganda:** Nikitin, *Unyielding City*, 17.

III. SPIES AND TRAITORS

41 **Knaak commanded the Eighth Company:** Paterson, *Hitler's Brandenburgers*, 147.

42 **"a company of saboteurs":** Höhne, *Canaris*, 377.

42 **"Oh, a long way back":** Jones, *Leningrad*, 20–21.

43 **twelve hundred planes:** Salisbury, *900 Days*, 106.

44 **a hundred miles into Soviet territory:** Ibid., 188.

44 **unit of thirty men:** Kurowski, *Brandenburger Commandos*, 109.

44 **a hot, dirty drive:** Gustav Klinter, as quoted in Jones, *Leningrad*, 19.

44 **demolition of the bridge:** *Kriegstagebuch 8. Panzer Division 26/6/41*, US National Archives, T315, Roll 484.

45 **botanist would entertain:** Golubev, *Great Sower*, 134.

45 **Ivanov had helped to transport:** Senin, *List'ya vyrastut vnov'*.

46 **thirty-minute commute:** Geyer, "Nine Hundred Days," 14.

47 **added to the second trainload:** Diary of Georgi Kniazev, July 7, as reproduced in Adamovich and Granin, *Leningrad under Siege*, 17, 19.

47 **"I saw them off":** Ibid., July 7 and 15.

47 **half a million books:** Tim Brinkhof, "The Creatures That Devoured Leningrad," History Today, May 28, 2020.

47 **Bureaucratic ineptitude:** In July, for example, children were sent to summer camps along the Luga River at Tolmachevo and Gatchina, directly in the path of the advancing German army, a failure that discouraged other parents from evacuating their children from the city in August.

48 **into the Leningrad province:** Dimbleby, *Barbarossa*, 182.

48 **"We're winning but the Germans are advancing":** Reid, *Leningrad*, 51.

48 **cut off by the German advance:** E.g., Adamovich and Granin, *Leningrad under Siege*, 19.

48 **four climbers concealed:** O. Oh. Matveyeva, "Chtoby Pomnili," website of St. Isaac's Cathedral State Memorial Museum.

49 **Alpine mountaineer named Olga Fersova:** Adamovich and Granin, *Leningrad under Siege*, 17.

49 **The Klodt stallions:** Salisbury, *900 Days*, 254.

49 **"a chattering fool":** Georgi Kniazev, diary entry August 11, 1941, as quoted in Adamovich and Granin, *Leningrad under Siege*, 35.

49 **death by firing squad:** In the first sixteen months of the war, nearly fifty-five hundred civilians were executed by the Soviet authorities for various offenses.

49 **"A spy for sure!":** Kochina, *Blockade Diary*, 33.

50 **Clothing considered too Western:** Jones, *Leningrad*, 87.

50 **even their home:** Reid, *Leningrad*, 125.

50 **city of collaborators:** According to Reid, there is no evidence of a genuine foreign spy operating in Leningrad during this period.

50 **Order 182:** A few days earlier, on June 27, the Leningrad city council issued an order mobilizing all able-bodied men between the ages of sixteen and fifty, and women aged sixteen to forty-five without young children, for civil defense work.

51 **erudition and gentleness:** Skrjabina, *Siege and Survival*, diary entry for July 18, 1941, 23.

51 **Ivanov had evaded dismissal:** Throughout the 1930s the Institute lost more employees than the total number of biologists who would eventually perish in Nazi Germany.

51 **report to Vyritsa:** Senin, *List'ya vyrastut vnov'*.

52 **"accustomed to the whine of shells":** Salisbury, *900 Days*, 197.

52 **helpers wore white headscarves and shawls:** Ibid., 383.

52 **"The soldiers marched quickly":** Olga Grechina, "Spasayus spasaya chast 1: Pogibelnaya zima (1941–1942)," *Neva* 1 (1994): 220–21.

53 **"There is no reason to be pessimistic":** Skrjabina, *Siege and Survival*, diary entry for July 18, 1941, 23.

53 **"the necessity of having to feed":** Halder War Journal, Evidence Division, Office of Chief of Counsel for War Crimes, Office of Military Government for Germany, 210.

53 **"a fatal event in the history of Europe":** Cameron and Steven, *Hitler's Table Talk*, 617.

53 **"annihilate the enemy completely":** Jones, *Leningrad*, 24.

IV. A TRAIN TO NOWHERE

56 **evacuees from Pskov:** Massie, *Pavlovsk*, 197.

57 **twelve hundred varieties of potato:** Senin, *List'ya vyrastut vnov'.*

58 **this would be the last trip:** Kameraz, in *V Osazhdennom Leningrade.*

59 **"There were always informants around":** As quoted in Jones, *Leningrad*, 88.

59 **"like leaving the twentieth century":** As quoted in Reid, *Leningrad*, 59.

59 **"we underestimated the Russian colossus":** August 10, 1941, Halder War Journal, Evidence Division, Office of Chief of Counsel for War Crimes, Office of Military Government for Germany (OMGUS).

60 **also for deserters:** Salisbury, *900 Days*, 210.

60 **executed by firing squad:** Reid, *Leningrad*, 72.

60 **tens of thousands of mines:** The commander of the German Fourth Panzers estimated his men had encountered a highly specific 26,588 mines.

61 **Ivanov helped the researcher V. F. Antropova:** Ivanov, "Sokhraneniye Kollektsii Kul'turnykh Rasteniy," 3.

61 **They received no training:** Simmons and Perlina, *Writing the Siege*, xii.

62 **citizens chased rumors:** Skrjabina, *Siege and Survival*, diary entry for August 25, 1941, 31.

63 **two standard carriages and an open-top freight car:** No. 194, Case No. 9, VIR.

63 **120 tons' worth of seeds:** Nikolai Ivanov, as quoted in Geyer, "Nine Hundred Days," 15.

63 **oversee the evacuation:** Isaak Izrailevich Prezent, DOB September 27, 1902.

63 **built-up heels and a tall green hat:** Ings, *Stalin and the Scientists*, 190.

63 **attempting to seduce:** This punishment did not, it seem, have any corrective effect. According to Simon Ings, when Prezent became chair of the Darwinian Departments at Moscow and Leningrad in 1948, he gave compulsory lectures in Michurinist biology. He would invite young female students to his personal apartment for the course exams. The young women soon advised one another to take a chaperone with them.

63 **"a very tight bunch of specialists":** As quoted in Kolchinsky, "Nikolai Vavilov in the Years," 344.

64 **"The successful Soviet writer":** Nabokov, *Speak, Memory*, 335.

64 **"Prezent works out the philosophy for me":** Lysenko, *Pod znamenem marksizma*, 186.

64 **install a fifth column:** Kolchinsky. "Nikolai Vavilov in the Years," 346.

64 **describe Vavilov's work as "racist":** Prezent, "Pseudoscientific Theories."

64 **scheduled to depart:** It is likely but not certain that the seed bank's staff traveled on the same train that pulled the freight carriage containing the seeds. The last passenger train departed from Leningrad the following day.

65 **"We'll have to work around the clock":** This dialogue is taken from the *Pravda* journalist Viktor Senin's short but historically useful 1979 book, *List'ya vyrastut vnov'*, an account based on interviews with Ivanov and other survivors. Dialogue should nevertheless be taken as impressionistic, rather than verbatim.

65 **no more than five pounds:** Information shared with author by Professor I. G. Loskutov, Vavilov Institute, September 2021.

65 **one of a hundred cereal grains:** Senin, *List'ya vyrastut vnov'*.

65 **"They want to uproot me, at my age?":** Ibid.

66 **offered his seat on the train:** Ibid.

67 **"knocked off the track by a giant hand":** Ivan Fedulov, as quoted in Jones, *Leningrad*, 99.

67 **tiny limbs caught in overheard wires:** When news of the attack reached Leningrad, officials in the city claimed the stories were no more than "hostile and provocative rumors."

68 **"Everything is split, charred, dead":** Inber, *Leningrad Diary*, entry for August 23, 1941, 10. Inber described the sights she saw as her train passed through Mga two days before the botanists. She was headed in the other direction, from Moscow to Leningrad.

69 **awoken by a sharp jolt:** The precise timings of the passenger train's journey remain somewhat unclear. Eichfeld passed directorship of the Institute to Virs on August 25, and the last passenger train left the city on the twenty-sixth. Mga was only lightly defended, and by the thirtieth the German army held the town through which the Institute employees were yet to pass. Not until September 8, 1941, did the seed bank's staff disembark, having returned to Leningrad. Eichfeld, it seemed, made it out of the city, presumably having taken an earlier train.

V. NO POTATO LEFT BEHIND

71 **thirty-six-year-old botanist:** Abram Yakovlevich Kameraz, DOB December 22, 1904, staff files, VIR.

72 **bombed the crowds of refugees:** Massie, *Pavlovsk*, 197.

72 **ordered everyone out to run for cover:** Kameraz, writing in Ivanov, *Kak my spasli mirovuyu kollektsiyu kartofelya*, *V blokadnom Leningrade*, 189.

72 **ground into a nearby ditch:** Reid, *Leningrad*, 148.

72 **smoldering peat bogs:** Burning of peat bogs reported by Georgi Kniazev, as quoted in Adamovich and Granin, *Book of the Blockade*, 256.

72 **carried an ultimatum:** Skrjabina, *Siege and Survival,* diary entry for September 8, 1941, 33.

72 **"Take every opportunity":** Salisbury, *900 Days*, 332.

73 **"Soviet Russia is kaput":** Ibid., 247.

73 **inspired a rumor:** Diary of Georgi Kniazev, as quoted in Adamovich and Granin, *Leningrad under Siege*, 36.

73 **The first shells landed:** Bidlack and Lomagin, *Leningrad Blockade*, 287.

73 **"We feel the Junkers circle":** Inber, *Leningrad Diary*, 13.

74 **"like the sky was falling down":** Mukhina, *Diary of Lena Mukhina*, 111–12.

74 **Explosion followed explosion:** During the attack, twenty-seven Junkers had dropped more than six thousand incendiary bombs across the city.

74 **the operetta continued:** Inber, *Leningrad Diary*, 16.

75 **"All hell was let loose in the sky":** Ibid.

75 **"It was so unlike smoke":** L. Shaporina, diary entry for September 8, 1941, as quoted in Simmons and Perlina, *Writing the Siege*, 23. Original held in the Manuscript Department of the Russian National Library.

75 **the monkeys in the zoo:** Sokolov, "Tyoplaya vanna dlya begemota," 153.

75 **unusually early frosts:** Simmons and Perlina, *Writing the Siege*, 107ff.

76 **Kameraz joined the seed bank:** Kameraz, autobiography, February 25, 1949.

76 **awarded a PhD:** Abram Yakovlevich Kameraz, staff file, VIR.

76 **Kameraz was responsible:** Kameraz did not accompany Vavilov on any specimen-gathering expeditions—Professor I. G. Loskutov to author, March 2022.

76 **six thousand varieties:** Kameraz, writing in Ivanov, *Kak my spasli mirovuyu kollektsiyu kartofelya, V blokadnom Leningrade*, 189.

77 **Kameraz disembarked:** Ibid., 189ff.

77 **a man named Korovichev:** Ibid.

77 **"Who cares about potatoes?":** Senin, *List'ya vyrastut vnov'.*

77 **feed the people of Russia:** Kameraz, writing in Ivanov, *V blokadnom Leningrade*, 189ff.

78 **his last moments:** Ibid. Kameraz describes these thoughts when taking shelter.

78 **Several hours:** Loskutov, "Wartime Activities."

79 **remain indoors during curfew:** Ivanov, "Sokhraneniye Kollektsii," 9.

79 **ten thousand firefighting units:** Salisbury, *900 Days*, 144.

79 **Shostakovich worked on a symphony:** Anderson, *Symphony for the City*, 202.

79 **firefighting group led by I. Ya. Yurtsev:** Orders 215 and 216, VIR.

80 **"And what if it is an explosive bomb?":** As quoted in Jones, *Leningrad*, 95.

80 **"Sheer hell":** Skrjabina, *Siege and Survival*, diary entry for September 8, 1941, 34.

81 **less than one month's supply:** Bidlack and Lomagin, *Leningrad Blockade*, 45.

81 **the city as an "island":** Peri, *War Within*, 5.

81 **"The enemy is at the gates":** Salisbury, *900 Days*, 304. In his later account of these events Kameraz did not specify the date on which he made his final rescue journey to Pavlovsk. He did, however, record that he was at the field station when the town fell to the German advance, at some point between September 9 and 16, 1941.

81 **"The dark and rusty contour of the barricades":** As quoted in Salisbury, *900 Days*, 285.

VI. CITY OF FIRE

83 **by the light of burning ropes:** Zherve et al., *Trudy Gosudarstvennogo Ermitazha*, vol. 8.

83 **made his way to the railway station:** In the *Pravda* journalist Viktor Senin's telling of this story, Kameraz treks the twenty miles to the Plant Institute on foot over several days. As per Kameraz's own brief recollection of the rescue, written ten years earlier, he returned to Pavlovsk Station after he regained consciousness. Kameraz does not explicitly state that he caught a train into the city, leaving some doubt as to whether he returned to the seed bank by foot or by rail.

84 **more than a metric tonne of potatoes:** Senin, *List'ya vyrastut vnov'*; and Kameraz, autobiography, February 25, 1949, VIR.

84 **mounted shelving onto the walls:** Borin, *Krutyye povoroty*, 17ff.

84 **The couple slept on the floor:** Kameraz, writing in Ivanov, *Kak my spasli mirovuyu kollektsiyu kartofelya, V blokadnom Leningrade*, 189.

84 **appointed new roles:** Order 287, VIR.

84 **divided into several subunits:** Order 289, VIR.

84 **Soldiers removed typewriters:** Ivanov, "Sokhraneniye Kollektsii," 9.

85 **fruit that bore his name:** Loskutov, "Wartime Activities."

85 **who had propagated perennial flowers:** Senin, *List'ya vyrastut vnov'*.

85 **held in a safe:** N. R. Ivanov's documents, 1945, Orders and Directives, VIR.

86 **landed on the roof:** In his unpublished memoir, Ivanov estimates that a total of one hundred incendiary bombs landed on the Institute roof.

86 **Believed to have superior hearing:** Eyewitness testimony recorded in Papernaya, *Podvig khraniteley*.

86 **with spots that looked like sprayed blood:** Varshavsky and Rest, *Saved for Humanity*, 120.

87 **"The metronome peacefully taps out seconds":** Natalia Uskova, diary entry, July 15, 1941, as reproduced in Peri, *War Within*, 24.

87 **monotonous routine:** All these recollections are drawn directly from Ginzburg, *Notes from the Blockade*.

87 **"How we hated the moon":** Olga Nikolaevna Grechina, oral history, as published in Simmons and Perlina, *Writing the Siege*, 104–15.

88 **"left feeling completely empty":** As quoted in Reid, *Leningrad*, 142.

88 **"burn our papers in the stoker":** Borin, *Krutyye povoroty*, 19ff.

88 **"short shrift to such faintheartedness":** Ivanov, in *V Osazhdennom Leningrade*, 188. In Alexander Borin's telling of this anecdote, he suggests that Ivanov took the stoker outside, filled it with scrap iron, and welded its door shut to prevent the burning of Institute valuables.

88 **small group of military personnel:** Generals Mikhail Khozin and Ivan Fedyuninsky.

89 **Forty-four:** Georgi Zhukov, DOB December 1, 1896.

89 **"defend Leningrad to the last man":** Zhukov, *Memoirs of Marshal Zhukov*, 300, 314–16.

89 **"shot on sight":** As quoted in Shukman, *Stalin's Generals*, 350. From September 12, 1941, small NKVD units known as blocking detachments were formed with the express purpose of shooting deserters.

89 **"responsible for the fall of Leningrad":** Reid, *Leningrad*, 126.

90 **boil from the thousand-degree heat:** Geyer, "Nine Hundred Days," 15–16; and Skrjabina, *Siege and Survival*, diary entry for January 19, 1942, 68.

90 **108 incendiaries:** According to Ivanov, five fires broke out on October 1, 1941, the night when the Institute sustained the greatest physical damage during the siege. In one of his many accounts, Ivanov named the staff members who put out the blazes that night as Bogdanov, P. A. Ivanova, I. I. Tumanov, and M. S. Belyaeva, with special mention of Golenishchev, Andreyeva, G. G. Tarasenko, Romanova, M. F. Volkova, and Yurtsev.

90 **in a suit, shirt, and tie:** Senin, *List'ya vyrastut vnov'*.

90 **"burned like a torch":** Boris Borisovich Piotrovsky, as quoted in Dimitri Ozerkov, "A Bomb Shelter for Passers-By," *Hermitage Magazine* 29 (December 2019): 103.

90 **"It's all one to me whether I live or die":** Georgi Kniazev, diary entry, September 29, 1941, as quoted in Adamovich and Granin, *Leningrad under Siege*, 65–66.

91 **"marked as necessary targets":** *Generaloberst* Alfred Joell, as quoted in *International Military Tribunal, Trials of War Criminals*, 15:413.

91 **"aimed at ruining the town"**: Sergeant Fritz Kopke, as quoted in ibid., 8:115.

91 **"it's my responsibility"**: Senin, *List'ya vyrastut vnov'*.

92 **no longer receive a salary**: *Spisok spetsialistov*, October 1941–January 1942.

92 **"gives way to the coldest reason"**: Cameron and Steven, *Hitler's Table Talk*, 44.

92 **"having to fee the populations"**: Halder War Journal, Evidence Division, Office of Chief of Counsel for War Crimes, Office of Military Government for Germany, 210.

93 **"embodiment of the infernal"**: As quoted in Jones, *Leningrad*, 16.

93 **"unprecedented, unmerciful and unrelenting harshness"**: Adolf Hitler, as quoted in Shirer, *Rise and Fall*, 830.

93 **"overcoming their personal scruples"**: International Military Tribunal, *Trials of War Criminals*, 11:516.

93 **"unheard-of harshness"**: Jones, *Leningrad*, 25–26.

94 **"starve the lot"**: Madajczyk, *Vom Generalplan Ost*, 440–42.

94 **"even a part of this city's population"**: International Military Tribunal, *Trials of War Criminals*, 1:58.

VII. FOUR OUNCES

97 **provided the only ventilation:** All details taken from conversation between Fillipovski and Mark Popovsky, December 9, 1968, as recorded in Popovsky, *Vavilov Affair*, 150. An unpublished letter to Popovsky, written by the academic A. L. Takhtadgian and found among Popovsky's papers in Columbia University's Rare Book & Manuscript Library records that the artist Fillipovski died in November 1987.

98 **summoned into Stalin's room:** As told to Dr. Yakushevsky by Vavilov and quoted in Loskutov, *Vavilov and His Institute*, 105.

99 **"hurry to produce immortal works":** Popovsky, *Vavilov Affair*, 123, and Golubev, *Great Sower*, 134.

100 **rarely sufficient to save one's life:** See Conquest, *Great Terror*, 128.

100 **"one-sided":** Popovsky, *Vavilov Affair*, 132.

100 **targets for further violence:** Testimony of the Russian playwright Vsevolod Meyerhold, who was executed a few months before Vavilov's arrest, as quoted in Shentalinsky, *KGB's Literary Archive*, 25–26.

100 **often produce a stranger:** As reported in the case of Isaak Rubin, one of the defendants in the 1931 Menshevik trial, when fourteen economists were tried and convicted of being members of a fictitious anti-Soviet secret organization. Rubin witnessed the murder of two strangers in this way on consecutive nights, before he negotiated a confession to stop the killings.

101 **depended on Vavilov's profession of guilt:** See Conquest, *Great Terror,* 131.

101 **wrote them up as statements:** Popovsky, *Vavilov Affair,* 133.

102 **to seven volumes:** Ibid., 135.

102 **shown to be forgeries:** Among the genuine denunciations against Vavilov was testimony from Aleksandr Karpovich Kol, head of the Plant Institute's department of plant introduction. Kol was, by all accounts, a careless worker who routinely lost or mislabeled seed specimens. After he was reprimanded by Vavilov for these errors, Kol seemingly plotted revenge against Vavilov, who was ten years his junior. In 1933 Kol made a deal with the secret police, providing his handlers with compromising material about Vavilov and other leading scientists at the Institute.

102 **indiscriminately burned Vavilov's papers:** Medvedev, *Unknown Stalin,* 67.

104 **personally chosen the panel's members:** Popovsky, *Vavilov Affair,* 154.

104 **including treason:** See also Conquest, *Great Terror,* 296.

105 **"After devoting thirty years":** Rokityansky, Vavilov, and Goncharov, *Sud Palacha,* 92.

105 **three main prisons:** The others were Lubyanka and Lefortovo.

105 **first person in the world:** V. S. Lekhnovich to Mark Popovsky, as quoted in Popovsky, *Vavilov Affair,* 99.

105 **seek an audience with Stalin:** Ibid., 147.

106 **sixty days' worth:** Pavlov, *Leningrad,* 49.

106 **Stalin refused to authorize:** Bidlack and Lomagin, *Leningrad Blockade,* 262.

106 **not to divert extra provisions to Leningrad:** Adamovich and Granin, *Book of the Blockade,* 348.

106 **"didn't have more foresight?":** Stadubsteva, *Krugovorot vremena I sudby,* 510.

106 **"squandered into private hands":** As quoted in Reid, *Leningrad,* 164.

107 **during the invasion of Poland:** Backe, the son of German émigrés, grew up in the southwestern corner of the Russian empire. During the First World War, he was arrested and interned in a village in the Ural Mountains. With the help of the Swedish consulate in Leningrad in 1918, Backe escaped Russia and fled to Germany, embittered by his and his family's experience. For Backe, the retributive policy toward Russia was both ideological and personal.

107 **"There must be absolute clarity about this":** As quoted in Tooze, *Wages of Destruction,* 480.

107 **given explicit instructions:** Ibid., 477.

108 **champagne, under the counter:** Pavlov, *Leningrad,* 51–53.

108 **eleven ounces:** Contrast this to the rations provided in the Japanese POW camps

on the River Kwai, where prisoners received a daily ration of 25 ounces of rice, 21 ounces of vegetables, 4 ounces of meat, 1 ounce of sugar, 1 ounce of salt, and 1 tablespoon of oil: a king's banquet by comparison.

109 **palatable dishes:** Simmons and Perlina, *Writing the Siege*, xli.

109 **"They are extremely unappetizing":** Skrjabina, *Siege and Survival*, diary entry, November 6, 1941, 45.

109 **"The enemy will die of starvation":** As quoted in Volkogonov, *Stalin*, 435.

110 **"tranquil there":** Kameraz, staff files, VIR.

110 **member of the Twenty-Third Army:** Ibid.

110 **assigned him to reconnaissance missions:** Kameraz, autobiography, February 25, 1949, VIR.

110 **"the skies will be blue":** As quoted in Salisbury, *900 Days*, 383.

111 **women volunteered:** Bidlack and Lomagin, *Leningrad Blockade*, 172.

111 **a specialist in botanical taxonomy:** Voskresenskaya, staff files, VIR.

112 **potatoes attracted attention:** Kameraz, autobiography, February 25, 1949, VIR.

112 **even provoke murder:** E.g., eyewitness testimony from Sofia Niklaevna Buriakova, as quoted in Simmons and Perlina, *Writing the Siege*, 98.

112 **two junior staff members:** Lekhnovich, staff files, VIR.

112 **"Those were my troops, indeed":** Pavel Gubchevsky, as quoted in Dimitri Ozerkov, "A Bomb Shelter for Passers-By," *Hermitage Magazine* 29 (December 2019): 103.

112 **the family evacuated:** Kuznetsova, staff files, VIR.

112 **one and only postcard home:** Paper by A. G. Gruzdeva, "In Memory of the Fallen of VIR," VIR.

113 **"His descriptions were so vivid":** Rod MacLeish, *The Hermitage: A Russian Odyssey* (Christian Science Publishing Society, 1994), vol. 3 in *From Czars to Commissars: A Museum Survives*.

113 **"fascinated by the chandeliers":** "How the Blockade Was Seen through the Windows of the Hermitage," interview with Oleg Chechin, *Rossiyskaya Gazeta*, January 2015.

114 **leaves of the *Cotinus*:** Johan Eichfeld, "The Wartime Work of VIR," *Research Report of the All-Union Plant Institute*, 1945, translated from the original Russian-language version provided by VIR staff.

114 **plants to dye fabric khaki:** According to VIR documents, Professor P. A. Yakovlev oversaw this work.

114 **extracted from food industry waste products:** Eichfeld, "Wartime Work of VIR."

114 **whistle cold through the corridors:** Ivanov, in *V Osazhdennom Leningrade*, 188.

114 **As a child:** Nikolai Rodionovich Ivanov, staff file, VIR.

115 **"Let's have a glass of hot water":** Senin, *List'ya vyrastut vnov'*.

VIII. CITY OF ICE

119 **"the whispering of the stars":** Thubron, *In Siberia*, 270.

119 **carried in his pocket a crude map:** Borin, *Krutyye povoroty*, 19ff.

120 **in the village of Dolzhok:** Biographical details courtesy of VIR.

120 **short stature, lean frame:** As described by Borin, "Podvig 13 leningradtsev."

121 **"It was difficult to walk":** As quoted in ibid.

122 **caused abdominal pain:** Bidlack and Lomagin, *Leningrad Blockade*, 264.

123 **"Why hadn't I bought it?":** Likhachev, *Reflections on the Russian Soul*, 220.

123 **a sort of pancake:** Reid, *Leningrad*, 182.

123 **"I am becoming an animal":** Berta Zlotnikova, as quoted in Dalya Alberge, "'Only Skeletons, Not People': Diaries Shed New Light on Siege of Leningrad," *Guardian*, December 24, 2016.

124 **"reach out for a pencil sharpener":** Ginzburg, *Notes from the Blockade*, 99.

124 **"the disappearance of my buttocks":** Boldyrev, *Osadnaya zapis*, diary entry for February 13, 1942, 58.

124 **Mothers divided their daily ration:** Testimony of Oleg Iatskevich, as quoted in Pimenova, *Deti voiny*, 130.

125 **"taking hours to eat":** Zyatkov et al., *Detskaya kniga voiny*, 18.

125 **"Lena screams and tears":** Kochina, *Blockade Diary*, 44.

125 **"Don't touch him!":** Yelena Razheva, oral history, Radio Free Europe, January 27, 2014.

125 **"wouldn't think twice":** Yura Riabinkin, diary entry, November 26, 1941, in Adamovich and Granin, *Leningrad under Siege*, 107.

126 **traders protested:** Salisbury, *900 Days*, 478.

126 **"corpses with the buttocks carved out":** As quoted in Simmons and Perlina, *Writing the Siege*, xxxi

126 **they all pleaded guilty:** Lomagin, *V Tiskakh Goloda*, 184–89. Other accounts of cannibalism in the beleaguered city exist, but even documentarians working decades after the events shied from including them in their published works. While collecting oral histories from siege survivors, Ales Adamovich and Daniil Granin were told about a mother and grandmother who used meat from one of the mother's dead children to feed the others, who were still alive. Such traumas were apparently too much for even the historians to include in their work.

126 **scavenging amputated limbs:** Reid, *Leningrad*, 289.

126 **"We've seen cases of kidnap":** Kochina, *Blockade Diary*, 72.

126 **arrested for cannibalism:** This, the lowest recorded figure of 1,380 arrests for alleged cannibalism, comes from Boris Belozerov in Barber and Dzeniskevich, *Life and Death*, 223.

126 **the primitive life of savages:** Alberge, "'Only Skeletons.'"

127 **for a pound of bread:** L. S. Rubanov file, Bakhmeteff Archive, Columbia University, 12–14, 19.

127 **"mind my own business":** Adamovich and Granin, *Leningrad under Siege*, 123.

127 **securing a replacement:** Salisbury, *900 Days*, 478.

127 **"would round up five people":** As quoted in Reid, *Leningrad*, 285.

127 **flint chinked against metal:** Ibid., 181.

128 **Bookish academics:** Ibid., 182.

128 **butchered their service animals:** By February 1942 only five police dogs were still in the service of the Leningrad police department.

128 **Pigeons disappeared:** This detail became an issue of some controversy after the war. The poet Anna Akhmatova described pigeons in the square in front of Kazan Cathedral. A survivor of the siege accused Akhmatova of ignorance or fabrication: the pigeons, she insisted, had all been eaten.

128 **"not knowing what cats and dogs were":** Salisbury, *900 Days*, 477.

129 **"the battle began to ease":** Lekhnovich, in *V Osazhdennom Leningrade*.

130 **bundles of around ten:** Loskutov, "*Wartime Activities.*"

130 **"need these seeds more than ever":** Ibid., 15.

130 **which also coated the metal boxes:** Ivanov, eyewitness report, as quoted in ibid.

130 **once-weekly inspections:** Ivanov, in *V Osazhdennom Leningrade*, 185ff.

130 **no evidence of any vermin:** Some believed that rodents abandoned the city en masse during the winter of 1941–42. According to the journalist Harrison Salisbury, Russian soldiers claimed that the vermin "made their way by the tens of thousands to the front-line trenches [where] food was more plentiful."

131 **trams began to run more infrequently:** Bidlack and Lomagin, *Leningrad Blockade*, 295.

131 **"just wouldn't come":** Ginzburg, *Notes from the Blockade*, 18.

131 **"our work saved us":** Borin, "Podvig 13 leningradtsev."

131 **said before his arrest:** As quoted by Alexanian, *N. I. Vavilov*, 3.

132 **murderous frosts that threatened the potatoes:** He was not alone in his concern. The damp and cold threatened many of the valuable objects stored for

safekeeping in the early weeks of war. In St. Isaac's Cathedral, across the square from the seed bank, around 120,000 paintings, fabrics, coins, books, sculptures, and pieces of porcelain had been brought and stowed from museums in the suburbs of Pushkin, Lomonosov, Petrodvorets, Gatchina, and Pavlovsk. The curators believed that the building's five-yard-thick walls would withstand direct artillery strikes. They had not accounted for the moisture, however, which was slowly absorbed by glue, causing antique furniture to bow and peel, and causing a freckling of corrosion on metal and bronze works.

133 **"multivolume edition of Tolstoy's works":** Matt Bivens, "New Facts Point Up Horror of Nazi Siege of Leningrad," *Los Angeles Times*, January 27, 1994.

133 **filled their boots with straw:** Suldin, *Blokada Leningrada*, 117.

133 **"fall below zero":** Lekhnovich, in *V Osazhdennom Leningrade*, 191ff.

133 **cotton wool, sacking, and rags:** Ibid.

133 **a couple of degrees:** Alexanyan and Krivchenko, "Vavilov Institute," 10ff.

134 **took the botanist an hour:** Geyer, "Nine Hundred Days," 14.

134 **onto a sheet of plywood:** Senin, *List'ya vyrastut vnov'*.

134 **half a cubic yard of firewood:** Lekhnovich, in *V Osazhdennom Leningrade*, 191ff.

134 **approached the elderly man:** Borin, *Krutyye povoroty*, 17ff.

134 **"Let me take the firewood":** Dialogue translated from Senin, *List'ya vyrastut vnov'*.

IX. A SILENCE OF ANGELS

137 **staff forlornly watched on:** Suldin, *Blokada Leningrada*, 119.

138 **lost fingers to frostbite or gangrene:** Dimbleby, *Barbarossa*, 446.

138 **Stalin's persuasive threat:** Shukman, *Stalin's Generals*, 351.

138 **key strategic points:** For example, the city of Rostov, on November 28, 1941.

138 **"self-confidence and self-belief":** As quoted in Jones, *Retreat*, 141.

139 **a single power plant:** Salisbury, *900 Days*, 417.

140 **"I had to stop and rest five times":** As quoted in ibid.

140 **"a way to pass the time":** Igor Kruglyakov, as quoted in Reid, *Leningrad*, 184.

140 **delirious with happiness:** Yevfrosinya Ivanovna, as quoted in Salisbury, *900 Days*, 418.

141 **not seen for several hours:** Some sources have reported Shchukin's date of death as December 27, 1941. The official staff file, seen by the author, unambiguously records the date as December 25, 1941.

141 **a packet of almonds:** All the scenic details here are from Senin, *List'ya vyrastut vnov'*, and Borin, *Krutyye povoroty*, 17ff.

141 **"personification of a hard worker"**: Staff files, 1942, 1962, VIR.

141 **"we are used to it"**: Diary of A. Vinokurov, January 9, 1941, Special Internal Correspondence No. 10042, NKVD USSR Administration for the Leningrad Region and the City of Leningrad (hereafter: FSB Archive), as published in Nikitin, *Unyielding City*, 9. Note, the author of the diary, A. Vinokurov, was arrested and executed by firing squad for his unvarnished observations and criticisms of the authorities' handling of the situation.

142 **"day lay all around"**: Ginzburg, *Notes from the Blockade*, 16.

142 **"Corpses in the gateways"**: Ibid., 29.

142 **frozen to death in the snow**: German Security Service report, as reproduced in Nikitin, *Unyielding City*, 21ff.

142 **"Bare, blue legs"**: Skrjabina, *Siege and Survival*, diary entry for November 28, 1941, 51.

142 **he was gone**: Ibid., diary entry for November 15, 1941, 49.

143 **of the uneaten bowl**: Diary of A. Vinokurov, FSB Archive, as quoted in Nikitin, *Unyielding City*.

143 **"to be buried is very difficult"**: Sorokin, *Leaves from a Russian Diary*, 229.

143 **via plane across Lake Ladoga**: Glantz, *Siege of Leningrad*, 67.

144 **"as I slowly die"**: Artist identified only as "ONOPKOV," FSB Archive, as reproduced in Nikitin, *Unyielding City*, 15.

144 **malnourishment and cold**: Nikitin, *Unyielding City*, 14, 16.

144 **disconnect their radio**: Skrjabina, *Siege and Survival*, diary entry for December 6, 1941, 54.

144 **caused their fuel to freeze**: Ibid., diary entry for January 19, 1942, 68.

144 **diary was confiscated**: Sevkabel Factory worker identified only as "Efimov," FSB Archive, as reproduced in Nikitin, *Unyielding City*, 17.

145 **ceased to menstruate**: Reid, *Leningrad*, 262.

145 **These women were all**: Simmons and Perlina, *Writing the Siege*, 32.

146 **"chink at ceiling level"**: Lekhnovich, in *V Osazhdennom Leningrade*.

147 **extinguish incendiaries**: All the information in this section is drawn from staff files, 1942, 1962, VIR.

147 **packets of rice samples**: Ibid.

147 **"It wasn't difficult"**: As quoted in Borin, "Podvig 13 Leningradtsev."

147 **On January 16, 1942**: Georgy Viktorovich Heintz, DOB December 23, 1894.

148 **"need those seeds more than ever"**: Geyer, "Nine Hundred Days," 15.

149 **"You'll never guess"**: Dialogue translated from Senin, *List'ya vyrastut vnov'*.

149 **sent from the front:** Brezhnev, who survived the war, became director of the Institute in 1965.

149 **easing away the nails:** Senin, *List'ya vyrastut vnov'*.

150 **rounded up the bodies:** Salisbury, *900 Days*, 434.

150 **lay in the bed for three weeks:** Skrjabina, *Siege and Survival*, diary entries for December 26, 1941, and January 13, 1942, 59, 66.

150 **ninety-eight orphanages:** Reid, *Leningrad*, 251.

150 **attempted to conduct experiments:** After the war Orbeli prepared a volume that listed more than a thousand scientific discoveries he claimed were made during the siege months. Although never published, one proof copy is held in the Academy of Sciences, another in the National Library of Russia in St. Petersburg.

151 **produce new works:** For example, Alexander Nikolsky, chief architect of the Hermitage, drew a series of sketches showing daily life in the museum.

151 **"the souls of the dead":** "How the Blockade Was Seen through the Windows of the Hermitage," interview with Oleg Chechin, 1988, reproduced in *Rossiyskaya Gazeta*, January 2015.

151 **"clenching its teeth":** Nikolai Markevich, diary entry for January 24, 1942.

152 **"You crazy woman!":** Senin, *List'ya vyrastut vnov'*.

153 **a report on staffing:** All figures are drawn from staffing report, June 1–December 31, 1941, VIR.

X. DO NOT FORGET MY NAME

155 **attentive dogs tugged:** Testimony of Dr. Andrei Sukhno, as quoted in Popovsky, *Vavilov Affair*, 176.

155 **returned to the kneeling position:** Pringle, *Murder of Nikolai Vavilov*, 269.

156 **"Spies! Traitors!":** Testimony of Dr. Andrei Sukhno, as quoted in Popovsky, *Vavilov Affair*, 176.

157 **Cell Block 3:** Ibid., 176ff.

158 **the administration told her:** Pringle, *Murder of Nikolai Vavilov*, 277.

159 **"paralyzed everything":** Cameron and Steven, *Hitler's Table Talk*, 200–201.

160 **"destroy the Russians for good":** Heinz Brücher letter to Victor Franz, March 11, 1942, courtesy of Dr. Uwe Hossfeld.

160 **youthful, outdoorsy experiences:** Daniel W. Gade, "Converging Ethnobiology and Ethnobiography: Cultivated Plants, Heinz Brücher, and Nazi Ideology," *Journal of Ethnobiology* 26, no. 1 (Spring/Summer 2006): 82–106.

161 **edited by Hitler himself:** Deichmann, *Biologists under Hitler*, 259.

161 **into an SS university:** Ibid.

161 **working with the Kaiser Wilhelm Society:** Brücher had previously taken a ten-month break from the army from September 1940 to June 1941 to complete a thesis titled "Die plasmatische Vererbung bei *Epilobium*"—"An experimental contribution to the genetic and developmental physiological significance of the cell plasma"—before heading to the Russian front.

162 **"impeccable soldierly behavior":** SS file on Heinz Brücher, as quoted in Uwe Hossfeld and Carl-Gustaf Thornström, "Retrospektive einer Botanikerflucht nach dem Ende des Zweiten Weltkrieges: Drei Briefe von Heinz Brücher an Erich Tschermak-Seysenegg (1948–1950)," *Haussknechtia* 10 (2004): 267–97.

162 **"The conquest in the east":** Heinz Brücher, "Die Wildrassen des Kaukasus und ihre Bedeutung für die deutsche Pflanzenzüchtung," *Der Biologe* 12, no. 4/5 (1943): 93–98.

163 **"attempt on the life":** All details of this interaction as told to Popovsky, *Vavilov Affair*, 177ff.

163 **little beard, and intelligent eyes:** Piotrovskaya's own description, as told to Mark Popovsky.

164 **"SS expedition to Tibet":** For more details of this expedition, see Meyer and Brysac's *Tournament of Shadows*.

164 **Brücher remained desperate:** Brücher was not alone in his concerns. At a conference meeting on February 9, 1942, Herbert Backe, state secretary for the Reich Ministry for Food and Agriculture met with high-ranking members of the SS and various divisions of the Kaiser Wilhelm Institute to discuss how the four of Vavilov's research stations currently within territory occupied by the Germans might be requisitioned to "save the valuable material and make further work there possible."

165 **"Leningrad is doomed to decay":** Cameron and Steven, *Hitler's Table Talk*, 400.

XI. THE ROAD OF GRACE

167 **stations still in operation:** Biographical details drawn from papers in VIR archives; the first evacuations of seed-bank staff began on February 17, 1942. It is unclear whether there were further evacuations of seed-bank staff throughout the spring before the ice thawed in April.

167 **chosen for evacuation:** Loskutov, "Wartime Activities."

167 **the last remaining, if meager, routes:** Zhdanov had allocated sixty-four planes to the route, but less than a third of this number were operational at any one time, delivering no more than fifty metric tonnes of food each day.

168 **using a horse-drawn sleigh:** Barber and Dzeniskevich, *Life and Death*, 55.

168 **157 trucks:** Jones, *Leningrad*, 220.

168 **tractors pulling snowplows:** As seen in a U.S. propaganda film showing the Road of Life, Department of Defense, Office of the Chief Signal Officer, ARC identifier 36071.

169 **bumper to bumper:** All details are drawn from oral histories with Vera Rogova and Nikolai Vavin, as recorded in Jones, *Leningrad*, 221ff.

169 **as many as three crossings a day:** Salisbury, *900 Days*, 414.

169 **"became a German colony":** Mikhail Tsekhanovsky, as quoted in Jones, *Leningrad*, 223.

170 **"I do not want to evacuate":** Varshavsky and Rest, *Saved for Humanity*, 160–61.

170 **disguised as cans of food:** Jones, *Leningrad*, 226.

170 **Order 25 arrived:** Courtesy of VIR staff.

171 **wrapped in blankets:** Senin, *List'ya vyrastut vnov'*.

171 **handed meal coupons:** Testimony of Vladimir Kulyabko, "Blokadniy dnevnik," as quoted in Reid, *Leningrad*, 276.

172 **trampled them:** Testimony of Tatiana Lassan, as published in Nikita Malsimov, "KOL na glinyanykh O," *Newsweek*, March 3, 2008.

172 **"Don't think ill of me":** Senin, *List'ya vyrastut vnov'*.

172 **providing a tour of the city's districts and sights:** In her diary entry for February 11, 1942, Elena Skrjabina, who took the train from Finland Station eleven days before the Plant Institute staff, mentions seeing these specific sights on her journey.

173 **especially rosy and healthy:** Skrjabina, *Siege and Survival*, 83.

173 **"blockade I never went hungry":** Pavel Luknitsky, as quoted in Jones, *Leningrad*, 226.

173 **"we had everything":** Ibid., 227.

173 **heat in the carriage:** Senin, *List'ya vyrastut vnov'*.

173 **Borisova Griva Railway Station:** Ivanov, in documents held at the VIR, records the distance his colleagues walked as seven kilometers, which correlates exactly to the distance between Borisova Griva Station and the "Broken Ring" monument on the Ladoga shoreline.

173 **huddled around bonfires:** Senin, *List'ya vyrastut vnov'*.

174 **gather up the corpses:** Reid, *Leningrad*, 274.

174 **but felt nothing:** It is possible this intimate detail, drawn from Senin's 1979 account, was passed to the author from a survivor of the journey, but it's likely to be a literary embellishment, albeit something that also probably took place.

175 **passed an overturned vehicle:** It is unclear whether this brief scene, which has Rubtsov witnessing the attack through a gap in the tarpaulin, was given as testimony to Senin from a survivor of the journey or merely included in his account as a scene emblematic of the German aerial attacks on trucks while they crossed the ice road.

175 **near the hamlet of Kobona:** Today a taxi ride from St. Petersburg to Kobona takes a little less than two hours.

175 **"hand them on":** This rather poetic dialogue, which forms just one part of a longer exchange, was almost certainly fabricated by Senin or his source to provide moral context and justification for Rubtsov's reported sacrifice. It is included here with that proviso. The precise timings of these events are also somewhat unclear. Senin asserts that Rubtsov died during the early leg of his journey, at a field hospital on Lake Ladoga's eastern shore. VIR documents mark the date of Rubtsov's death as 1400 April 14, 1942. Either this date is an estimate, Senin's description of the circumstances of Rubtsov's death questionable, or, if these are the facts, then Rubtsov evacuated from the city a few weeks after his colleagues, who departed the city in mid-February.

176 **less than a hundred miles:** Bidlack and Lomagin, *Leningrad Blockade*, 272.

176 **pushed his corpse out:** Testimony of Vladimir Kulyabko, as recounted in Reid, *Leningrad*, 277.

177 **indifferent to such judgment:** Kochina, *Blockade Diary*, 108.

177 **"irresponsible and heartless":** Report to the Leningrad oblast Party committee, March 5, 1942, Dzeniskevich, *Leningrad v osade*, doc. 137, 292–94.

177 **tied to his chest:** Loskutov, "Wartime Activities."

177 **just fourteen:** These figures derive from two sources. Ivanov recorded that "around thirty people occupied the building as a kind of barracks." Four years later, Ivanov participated in an English-language article for *Wildlife* magazine, which claimed that, by the spring of 1942, that figure had reduced to fourteen. This number does not, however, include building staff, caretakers, and laborers; in total, according to the various primary source documents, no more than nineteen individuals were spread across the two buildings following the evacuation.

177 **use of a walking stick:** Geyer, "Nine Hundred Days," 14ff.

177 **no aerial bombardment:** In January 1942 the 2,696 shells fired at Leningrad by German artillery guns was half the number that had rained onto the city the previous month. It was a short reprieve. In February 4,776 shells exploded within the city, and 7,380 in March.

178 **four other museums:** Varshavsky and Rest, *Saved for Humanity*, 192ff.

178 **the thief had taken:** Ivanov, in *V Osazhdennom Leningrade*, 185ff.

178 **four-ounce increase:** All figures from Appendix A, "Daily Bread Rations," Bidlack and Lomagin, *Leningrad Blockade*, 413.

179 **including a selection of potatoes:** Borin, *Krutyye povoroty*, 18ff.

179 **a full duplicate:** Senin, *List'ya vyrastut vnov'*; and Loskutov, "Wartime Activities."

179 **received a telegram:** The telegram was addressed to K. A. Panteleyeva and G. N. Reuter.

180 **"we'll be held to account":** As quoted in Senin, *List'ya vyrastut vnov'*.

181 **"of secondary importance":** Ibid.

181 **"small group of emaciated people":** Ivanov, "Sokhraneniye Kollektsii," 12.

XII. CITY OF GARDENS

184 **"of minor importance":** Senin, *List'ya vyrastut vnov'*.

184 **thirty pea sprouts per day, per person:** Ivanov, undated, unpublished manuscript, 5, VIR.

185 **"we must continue":** This dialogue, which should be taken as impressionistic rather than verbatim, is translated from Senin, *List'ya vyrastut vnov'*.

185 **reduced by up to a half:** Reid, *Leningrad*, 341.

185 **"now we could see it":** Vladimir Garshin, as quoted in ibid.

185 **dried out dampened books:** Loskutov, "Wartime Activities."

186 **to sit and take a break:** Senin, *List'ya vyrastut vnov'*.

186 **staff recorded their work:** VIR.

186 **"What's all the commotion?":** Ibid.

187 **"We're starving to death":** Senin records this scene in his short book. He does not identify the exasperated speaker, and the dialogue is almost certainly fabricated. But undoubtedly there were such debates among the starving staff as to whether they had the right to eat expired samples.

187 **"That's the rule":** Senin, *List'ya vyrastut vnov'*.

188 **"child's play":** Tikhonov, *Defence of Leningrad*, 39.

188 **a program to vaccinate:** Salisbury, *900 Days*, 507.

188 **"Let's protect ourselves":** Senin, *List'ya vyrastut vnov'*.

189 **"Let them feed us first":** Reports to Zhdanov from Antyufeyev, head of the "instructors" department of the City Party Committee, TsGAIPD SPb: Fond 24, op. 2v, delo 5760.

189 **"beautiful as a flower-strewn glade":** Inber, *Leningrad Diary*, 734.

189 **"excavation of some ancient city":** Olga Grechina, as quoted in Reid, *Leningrad*, 343.

189 **hand-rolled from torn scraps:** Salisbury, *900 Days*, 5.

189 **"crack it with a pick":** As quoted in Reid, *Leningrad*, 344.

189 **meltwater flowed in eager streams:** "Chtoby pomnili . . . ," https://cathedral.ru /ru/isaac/memory, retrieved March 2023.

190 **tally of the dead:** Figures recorded by the Leningrad Funeral Trust.

191 **consigned to a purgatory:** Salisbury, *900 Days*, 510.

191 **"I rode the streetcar!":** Ibid.

191 **She had moved into:** Pringle, *Murder of Nikolai Vavilov*, 284.

191 **'The city again is lively':** As quoted in Salisbury, *900 Days*, 510.

191 **"just in empty space":** Olga Bergholz diary entry for May 13, 1942.

192 **"I can't feel anything at all":,** Mukhina, *Diary of Lena Mukhina*, diary entry for May 25, 1942, 389.

192 **"The great sufferings of winter":** Ginzburg, *Notes from the Blockade*, 101.

192 **"cruel, dishonorable, humiliating acts":** Ibid., 77.

192 **"What an immense range":** As quoted in Adamovich and Granin, *Leningrad under Siege*, 194.

193 **"The enemy wanted to kill the city":** Ginzburg, *Notes from the Blockade*, 100.

193 **eat raw on the spot:** Senin, *List'ya vyrastut vnov'*.

194 **medicine to treat scurvy:** Ibid.

194 **died from eating poisonous varieties:** Dzeniskevich, *Leningrad v osade*, doc. 147, 312.

195 **relayed via posters:** Salisbury, *900 Days*, 535.

195 **his mother's collection:** Senin, *List'ya vyrastut vnov'*.

196 **"Vegetable plots are springing up":** Tikhonov, *Defence of Leningrad*, 42.

196 **seven square miles of private gardens:** Bidlack, *Workers at War*, 28.

196 **providing growing advice:** N. R. Ivanov's documents, 1945, VIR.

196 **"Fifteen-hundredths":** Lazarev, *Vospominaniya o blockade*, 211–12.

197 **"People don't trust one another":** Petrovskaya Wayne, *Shurik*, 186.

199 **Panteleyeva and Kordon remained:** N. R. Ivanov's documents, 1945, VIR.

199 **Maria Belyaeva, sat watch:** Loskutov, "Wartime Activities."

199 **"Tears are proof":** Zherve et al., *Trudy Gosudarstvennogo Ermitazha*, 8:79.

200 **handfuls of green grass:** Luknitsky, *Sobraniye Sochinenii*, as recounted in Salisbury, *900 Days*, 521.

200 **"not yet completely cleared":** As quoted in Jones, *Leningrad*, 252.

200 **"Yurka":** Bergholz's nickname for Yuri Makogonenko, her colleague at Radio

House, who became her third husband. Makogonenko temporarily lost his job for allowing the broadcast of a banned poem, Zinaida Shisova's "Road of Life," which referred to a corpse stored on a balcony. Following a call from the Party Committee, the recital was halted midstanza.

200 **"reborn in me":** Bergholz, *Govorit Leningrad*, diary entries for May 13 and June 3, 1942.

200 **punched through the wreckage:** Tikhonov, *Defence of Leningrad*, 48–49.

XIII. BEAT, BEAT, AND ONCE AGAIN BEAT

204 **as high as 30°C:** Eyewitness testimony of Angelina Rohr, a doctor and journalist, as told to Popovsky, *Vavilov Affair*, 178.

205 **three or four men squeezed:** Ibid.

206 **elected president of the International Congress:** Karl Saks, "Soviet Biology," *Science*, December 21, 1945.

206 **on the immunity of plants:** Kolchinsky, Hossfeld, and Levit, "Russian Scientists."

206 **twenty scientists signed a proposal:** Rokityansky, Vavilov, and Goncharov, *Sud Palacha*, 106.

207 **Britain's prestigious academy of sciences:** The Royal Society was not the only British establishment to show Vavilov solidarity in this way. In 1942 the Russian scientist was also elected a member of the Geographical Society of London, the Linnaean Society of London, and the Royal Society of Edinburgh.

207 **sent the form to Moscow:** "Vladimir Komarov, President of the USSR Academy of Sciences," *Nature* 154 (1944): 634–35.

207 **"We expected the signature of Nikolai":** Yuri Vavilov archive, as quoted in Pringle, *Murder of Nikolai Vavilov*, 275.

208 **punching and biting:** Testimony of Nesvitsky, as told to Popovsky. In the notes to his book *The Vavilov Affair*, Popovsky points out that placing a vulnerable but violent individual into a small cell with political prisoners was "not a new trick." He quotes from a 1920 magazine, *Hard Labor and Exile*, which includes reports of tsarist gendarmes treating revolutionaries in the same way, a practice known as contenting.

208 **"These people were anonymous":** Testimony from Irina Reznikova, *Repressii v period blokady Leningrada*, as quoted in Reid, *Leningrad*, 306.

209 **"unfit for work":** Petition from the Corrective-Labor Camps and Columns Directorate of the NKVD to State Defense Committee emissary D. V. Pavlov, December 31, 1941, in Dzeniskevich, *Leningrad v osade*, doc. 175, 413.

209 **"we will shoot you":** Popovsky, *Vavilov Affair*, 182.

210 **"even at the lowest level":** Letter held in Vavilov's police records, as published in ibid., 183.

210 **Pryanishnikov had nominated Vavilov:** Ings, *Stalin and the Scientists*, 300; and Conquest, *Great Terror*, 296.

211 **"a period of twenty years each":** Letter no. 52/8996, June 13, 1942, investigation case no. 1,500, vol. 1, NKVD.

211 **"genetical investigations":** Certificates of election and candidature for Fellowship of the Royal Society, TRS, EC/1942/24.

212 **onions and tobacco:** Popovsky, *Vavilov Affair*, 186.

212 **carried him outside:** Medvedev, *Rise and Fall*, 74.

212 **"All the time we feel for Stalingrad":** As quoted in Jones, *Leningrad*, 269.

213 **"ripped open a hornet's nest":** Alexei Amelichev, as quoted in ibid., 270.

214 **"the musicians in our artillery orchestra":** As quoted in ibid., 272.

215 **returned to the prison hospital:** Identified by Major Vasili Andreyev as Vavilov's last refuge.

216 **"What on earth are you asking for?":** Testimony of Viktor Shiffer, a mining engineer and former prisoner at Saratov, as told to Popovsky, *Vavilov Affair*, 188.

216 **"Corpse of a man":** Report by Zoya Rezayeva, as quoted in ibid., 191.

216 **drink instead:** "Who would waste spirits on his hands?" Novichkov told Popovsky in 1967.

217 **knocked a metal spike into the dirt:** Novichkov told Popovsky that in January 1943 he was told one of the men he was due to bury was a famous person. This corpse wore clean underwear and was to be buried in a wooden box—the only time, Novichkov claimed, he had seen such an honor. Popovsky decided not to investigate the claim further. It is not so important, he wrote, "to know whether the bones of a long-suffering academician rotted in one hole or another." More important, he added that "posterity should not forget the terrible sufferings that Russia's intellectuals went through in Stalin's torture chambers."

XIV. A FARM ON THE FRONT LINE

219 **broken free of their boxes:** This account is reported in Jones, *Leningrad*, 281.

220 **reach those in need:** Simmons and Perlina, *Writing the Siege*, xxiii.

220 **raced to fix the tracks:** Repair crews patched the track more than twelve hundred times before the Germans were pushed back from their position on the Sinyavino ridge in September 1943.

220 **Praskovya Petrova, and Anna Andreyeva:** Ivanov, in *V Osazhdennom Leningrade*; and Senin, *List'ya vyrastut vnov'*.

220 **Institute for the Predportovy:** Ibid. Senin, *List'ya vyrastut vnov'*.

220 **only a mile:** Ivanov, "Sokhraneniye Kollektsii."

221 **"the devil's bridge":** Salisbury, *900 Days*, 555.

221 **he showed his papers:** Senin, *List'ya vyrastut vnov'*.

222 **insufficient to feed the nation:** Bidlack and Lomagin, *Leningrad Blockade*, 287.

222 **"Those are German positions":** The dialogue is translated from Senin, *List'ya vyrastut vnov'*.

222 **four women wearing harnesses:** Ibid.

223 **A searchlight ignited:** Ibid.

223 **"Get down":** As quoted in ibid.

224 **protected by a magic spell:** It's unclear why the plowers were allowed to work during the daytime, when it was considered too dangerous for the botanists to work in a similar way.

225 **"shrinking in the sunlight":** As quoted in Jones, *Leningrad*, 279–80.

225 **failed to take:** Lekhnovich and his team replaced the dead potatoes with turnips.

226 **scheme had been a success:** According to documents in the VIR archives, more than twenty million potato seedlings were cultivated in the city between 1941 and 1946, thanks in major part to the methods of accelerated propagation developed and taught by Lekhnovich and Voskresenskaya.

227 **she could walk again:** Jones, *Leningrad*, 280–81.

227 **published a monograph:** Olga Voskresenskaya, staff files, courtesy of VIR staff.

227 **almost completely blind:** This information comes from Voskresenskaya staff file; Senin, in his account, writes that she lost her vision because of malnutrition and makes no reference to the head injury that would later claim her life.

227 **their spoken accounts:** According to the VIR staff files, Voskresenskaya's mentor and tutor S. M. Bukasov could, even into old age, identify many potato species by smell alone.

227 **fragile semblance of normality:** In July 1943, for example, 210 were killed and 921 wounded by German shells.

227 **continued to provide support:** The city's total vegetable harvest in 1943 was more than twice as large as in the previous year, despite a smaller population.

228 **patches of potatoes and turnips:** Salisbury, *900 Days*, 555.

228 **"just a hindrance":** Vsevolod Vishnevsky, as quoted in ibid., 557.

228 **"the echo comes in layers":** Inber, *Leningrad Diary*, diary entry for February 9, 1943, 133.

228 **"one wants so much to live":** Ibid., diary entry for April 16, 1943, 140.

228 **"calm, lively, and jolly"**: Eyewitness testimony of Nikolai Chukovsky, as reproduced in Salisbury, *900 Days*, 555.

228 **Never in history**: Ibid., 556.

229 **"The position is unsuitable"**: Jones, *Leningrad*, 280.

229 **"as beautiful as ever"**: Inber, *Leningrad Diary*, diary entry for July 17, 1943, 155.

230 **cigarette still lodged**: Jones, *Leningrad*, 281–82.

231 **"It's not only sweat"**: Senin, *List'ya vyrastut vnov'*.

XV. THE SEED COMMANDO

233 **filled with treasure**: *Bericht über das Sammelkommando*.

234 **first raised the idea**: Nikita Malsimov, "Kolos: Na Glinyanykh Nogakh," *Newsweek*, March 3, 2008.

234 **"all the more significant"**: Ibid.

235 **through military fiasco**: Hossfeld and Thornström, "Instant Appropriation."

235 **Brücher met with**: There exists a potential discrepancy in Brücher's correspondence regarding the date of his meeting with Pohl. In his letter to the SS Office for Ancestral Heritage, dated April 14, 1944, he writes, "On the order of *Obergruppenführer* Pohl of 8.5.43 I had the task of seeking out agricultural and scientific propagation research stations in the area of higher SS and Police Command, Southern Russia." In his report on the SS Collection Commando, he writes, "I therefore submitted on 1.6.43 . . . the suggestion that we should send a collection command force into the occupied eastern zones."

236 **state-organized biopiracy**: The term *biopiracy* was introduced to the Convention of Biological Diversity in December 1993.

237 **can of Olivier salad**: Senin, *List'ya vyrastut vnov'*.

237 **visited the farm**: Ivanov, "Sokhraneniye Kollektsii."

238 **an overall success**: Ivanov, in *V Osazhdennom Leningrade*.

239 **built a mill**: According to VIR documents, Prokofyev built the mill in 1922. Known as Prokofyev's mill, it was still used at the Pushkin laboratory into the 1970s. In 1941, shortly before he died, Prokofyev built a second, larger mill for buckwheat.

239 **oversaw the establishment**: All detailed information on VIR experimental stations is from Alexanian, *N. I. Vavilov*, 8–9.

240 **of the glamorous Pushkin Station**: After the Germans occupied Pushkin, Dr. Walter Hertzsch, head of the Kaiser Wilhelm Institute's branch for plant breeding in East Prussia, arrived at the manor to supervise the research.

240 **the Polar station**: *Research Report of the All-Union Plant Institute*, 1945, courtesy of VIR staff.

240 **Brücher introduced himself:** Email to Professor Uwe Hossfeld from Arnold Steinbrecher, June 13, 2007, courtesy of Hossfeld.

241 **albeit often neglected:** Professor Uwe Hossfeld argues that Brücher had overestimated the value of the plant material held in the *ex situ* field stations, away from the main collection in Leningrad, as the rise of Lysenkoism had broadly halted meaningful experimentation in these locations during the late thirties and early forties.

242 **accusations of collaborating:** See, for example, Elina, "From Russia with Seeds."

243 **especially galling:** Letter from Brücher to the personal staff, RF-SS Office for Ancestral Heritage, April 14, 1944, Bl. 60/61, BDC.

243 **"no seeds were to be handed over":** Ibid.

244 **refused to part with seeds:** Ibid.

244 **"You bear full responsibility":** When he returned to Lannach Castle, Brücher did not put the matter to rest. He passed his report to the military administration office hoping that the Gestapo might investigate what he considered to be a clear case of insubordination (in addition to an attack on his ego). Brücher's desire for vengeance was not sated. After investigating the matter, a representative for the SS concluded that Brücher's "attacks" were unwarranted, that he had not been in possession of the full facts, and that, in the view of the scientists whose reputation he had impugned, his report was "careless and overbearing," the work of "a youthful hothead."

245 **a fortified rival:** Following the publication of a 2008 *New Scientist* article that included information about the seed commando, the magazine received an anonymous letter from a well-informed reader who claimed, "The sum total of samples collected in expeditions within the Soviet Union would have filled two backpacks." The extensive planting schedule at Lannach Castle suggests this was not the case, and during its mission across several weeks, visiting eighteen experimental stations, the unit harvested many seeds and samples.

XVI. TWENTY-FOUR SALVOS

247 **pardon for Sergei's brother:** Interview with Maria Shebalina, as quoted in Popovsky, *Vavilov Affair*, 163–64.

248 **"baked blood":** Diary entries as quoted in Pringle, *Murder of Nikolai Vavilov*, 277.

249 **"Died, January 26, 1943":** S. I. Vavilov, diary entry, July 5, 1943.

249 **William Denton Venables:** The "Denton" was a nugget of personal aggrandizement; William Venables was his given name.

249 **plaster had begun to break away:** Nordqvist, *Den Stora Frostolden*, 242ff. Nearly forty years earlier Franz Kandler, a twenty-seven-year-old businessman, assumed management of the castle and its surrounding estate. Kandler ran the Lannacher

roof-tile and pottery factory on the estate before he moved into local politics in 1927. When the Nazis annexed Austria, Kandler began leasing the castle to the SS. A small force moved in on May 14, 1938, and used the grounds to train Austrian police. Two months later the SS ordered Kandler to sell them the property. When he refused, they impounded the building, which went some way to explain the dilapidation.

250 **lose his pursuers among the vines:** Liberation Report: Trooper W Denton Venables (service number 7911803), TNA WO 208/3336/257.

251 **the POW met:** The precise date when Venables arrived at Lannach remains unclear but is likely to have been in the late autumn 1943.

251 **higher elevations and in colder climates:** Arnold Steinbrecher, as quoted in Pearce, "Great Seed Blitzkrieg," 38–41.

251 **added to the wheat:** Approximately sixteen hundred varieties of barley, seven hundred varieties of wheat, and seven hundred varieties of oats were harvested during the Tibetan mission. For further details see Meyer and Brysac, *Tournament of Shadows*.

252 **fast-ripening varieties:** Nordqvist, *Den Stora Frostolden*, 242ff.

252 **grow in the arctic:** Arnold Steinbrecher, as quoted in Pearce, "Great Seed Blitzkrieg," 38–41.

252 **spoke fluent German:** Liberation Report: Trooper W Denton Venable.

252 **twenty-eight years old:** Heinz Brücher, DOB January 14, 1915; William Denton Venables, DOB April 10, 1915.

252 **began to collaborate:** Arnold Steinbrecher, as quoted in Pearce, "Great Seed Blitzkrieg," 38–41.

253 **"my British wartime collaborator":** Ibid.

253 **"a quiet, fast, invisible death":** S. I. Vavilov, diary entry, July 6, 1943.

253 **"like a person who is somewhere else":** As quoted in Pringle, *Murder of Nikolai Vavilov*, 282.

253 **"The last thin thread":** S. I. Vavilov, diary entry, October 26, 1943.

254 **as part of Operation Iskra:** Abram Kameraz, staff files, courtesy of VIR; and Glantz, *Siege of Leningrad*, Appendix 1: "Red Army Order of Battle in the Leningrad Region," 199–211.

254 **"kept a great secret":** Reid, *Leningrad*, 382.

255 **stenciled inscriptions:** Jones, *Leningrad*, 284.

256 **"We're living through hell":** As quoted in Reid, *Leningrad*, 384.

256 **crashed to the ground:** Jones, *Leningrad*, 284.

256 **"all will be well":** Inber, *Leningrad Diary*, 179.

256 **"not flowing in vain"**: Ibid.

257 **"the strength of the flames"**: Ibid., 180.

257 **"almost human"**: Ibid., 181.

258 **struggling to keep up**: Salisbury, *900 Days*, 566.

258 **"Didn't make it"**: Reid, *Leningrad*, 395.

XVII. CASTLE OF SPOILS

261 **burned or cut down**: Eyewitness details from Vera Inber's journey to Leningrad from Moscow, undertaken shortly before Sergei Vavilov's own journey.

262 **yet to return**: Evacuees began to return in the summer of 1944, more than doubling Leningrad's population in twelve months.

262 **inconsolably sobbing**: Oral history of Irina Ivanova, née Bogdanova, as quoted in Reid, *Leningrad*, 391.

262 **"For what purpose?"**: Olga Grechina, "Spasayus spasaya chast 2: Skazka o gorokhovom derive (1942–1944 gg.)," *Neva* 2 (1994): 246.

262 **649,000 had succumbed to starvation**: This was the figure cited by the Soviet government at the Nuremberg war crimes trials. Most historians agree that it represents a substantial underreporting, as many deaths were never registered. Neither do these figures include deaths in rural areas within the siege ring, nor the thousands who died while evacuating from the city, either on the ice road or on the trains they boarded thereafter.

262 **"not the only place to suffer"**: As quoted in Volkogonov, *Stalin*, 435–36.

263 **"If only God and souls"**: S. I. Vavilov, diary entry, May 20, 1944.

264 **a showroom of battered tanks**: Details drawn from Vera Inber's eyewitness account, *Leningrad Diary*, 204ff.

265 **"the ghosts of our dead"**: As quoted in Jones, *Leningrad*, 290.

266 **These nine women**: Their names were Josepha Adamski, Anna Ambrozkova, Ida Bartosch, Martha Daig, Anna Fleischmann, Margarethe Hempel, Luise Schramm, Maria Schwenk, and Georgijne Verich.

266 **Their faith . . . forbade them**: Arnold Steinbrecher, as quoted in Pearce, "Great Seed Blitzkrieg," 38–41.

267 **"sow in the mountains this spring"**: BA Berlin, SS-Akte Brücher, BDC.

268 **"permission to visit"**: Liberation Report: Trooper W Denton Venables (service number 7911803), TNA WO 208/3336/257.

270 **unwilling to move out**: Many regular servicemen did not automatically receive rights to their prewar accommodation.

270 **promoted to the rank of sergeant major:** Kameraz autobiography, February 25, 1949, courtesy of VIR.

270 **Nor had the couple's baby:** Kameraz, who wrote two brief memoirs of his wartime experience, did not record the circumstances in which he received this information, nor the effect it had on him. The omission of this information suggests something of the extent of the psychological and emotional damage it caused him.

270 **Kameraz returned:** In May 1948 Kameraz became director of the laboratory for potato selection. He published a 360-page book, coauthored by S. M. Bukasov, in which the pair outlined the results of their research. Through this work he produced valuable varieties such as the Kameraz 1, Detskoselsky, Pushinsky, and Volkhovsky, and eventually the cancer-resistant cross-species hybrid Rubin, a variety still grown and sold around the world.

271 **eager to deflect blame:** Medvedev, *Unknown Stalin*, 206. Lysenko, whose name was also among the list of twenty-two candidates sent to the Kremlin to be considered for promotion, was discounted.

272 **future star mathematician:** Details contained in NKVD files held in the State Archive of the Russian Federation, discovered by Yuri Vavilov and G. Rokityanskiy. As reported by Olga Boguslavskaya, "Dropped on Orders from Above," dyatlovpass.com.

272 **into the abyss below:** Ibid.

274 **four hundred photographs:** Senin, *List'ya vyrastut vnov'*.

274 **"eaten by the famished people":** S. C. Harland and C. D. Darlington, "Obituaries, Prof. N. I. Vavilov," *Nature* 3969 (November 24, 1945).

275 **from a BBC broadcast:** Borin, *Krutyye povoroty*, 20ff.

275 **Nobody . . . could have believed:** Nikolai Ivanov, as quoted in Geyer, "Nine Hundred Days," 15.

POSTSCRIPT

277 **consumed by vermin or failed to germinate:** Ivanov papers, 1945, courtesy of VIR staff.

277 **sixty-six thousand varieties of flowers:** Ibid.

278 **used across the world:** Golubev, *Great Sower*, 140.

278 **disease-resistant variety of wheat:** Ibid., 145.

278 **a favored crop of farmers:** Staff files, courtesy of VIR.

278 **a third of all Russia's arable land:** Senin, *List'ya vyrastut vnov'*.

278 **bred from original samples:** "Genetic Resources Departments," information

pamphlet produced by VIR. The collection currently contains more than 40,000 accessions of wheat from more than seventy countries; 37,000 varieties of oat, rye, and barley; 160 species of grain legume, including 8,000 accessions of pea; 7,000 accessions of common bean; 3,000 lentils; 49,000 accessions of varieties of maize, sorghum, buckwheat, rice, and millet; 22,000 accessions of fruit—apple, pear, quince, sweet cherry, plum, apricot, peach, and grape; and more than 280 species of vegetables.

278 **found in no other scientific collections:** "The Battle to Save Russia's Pavlovsk Seed Bank," *Guardian*, September 20, 2010.

279 **"two-tongued duplicitous literature":** Foreword to Barskova, *Living Pictures*, xiv.

280 **the restoration of his brother's reputation:** Trofim Lysenko remained the dominating figure in Soviet biology until 1964, when his theories were finally discredited, and he was deposed from his position (while retaining his prestigious title, generous salary, and other face-saving privileges at the Academy of Sciences). A few years later Loren Graham, professor of the history of science at MIT, met Lysenko, whom Graham described as a "gaunt and homely man," in the dining room of the House of Scientists in downtown Moscow. As the pair ate together, Lysenko denied the accusation that he had been responsible for the deaths of Russian biologists. "I disagreed with Vavilov on biological issues," he said, "but I had nothing to do with his death in prison. . . . I am not responsible for what . . . the secret police did in biology." Graham, who had for months been studying Lysenko's work and victims in Soviet libraries and archives, was unconvinced. "His method," Graham wrote, "was lethal and passive-aggressive." By accusing Vavilov and others of being traitors to the Soviet cause, even foreign agents harming Soviet agriculture, Lysenko had used his power and position to unleash the secret police on his critics. Even if his hand did not lock the cell door, Lysenko's crucial denunciations (*donos*, as this method of using accusation as a punitive tool to remove political, professional, or personal rivals was known) implicated him in Vavilov's fate. With the full and terrible force of state power at his disposal, this humble peasant outsider became complicit in the murder of those he saw as snobbish aristocrats.

280 **"The authorities had no time for it":** Pavlov, *Leningrad*, 129.

282 **"I saw people in tears":** Popovsky, *Vavilov Affair*, 8.

283 **"extended as far as the moon":** "Moon Crater Named for Once-Disgraced Soviet Geneticist and Brother," *New York Times*, December 3, 1968.

283 **interviewing more than a hundred:** Letter from Mark Popovsky to Alexander Chakovsky, chief editor of *Literaturnaya Gazette*, January 8, 1987, 4078435, CUL.

284 **collective, flawless sacrifice:** The sons and daughters of those who survived the siege were often those most eager to protect the conventional Soviet narrative of a city

population roused to unblemished heroism by unrivaled hardship. Like the beaches of Dunkirk to the British baby boomers, the city of Leningrad became a defining redemption story, illustrative of a fleeting brand of heroism embodied by their parents' and grandparents' generation. "Actual *blokadniki*," the journalist Anna Reid wrote in her 2011 book, *Leningrad*, "are anxious to stress the siege experience's closed, stony quality; its complete lack of redemptive value and the depth of damage done."

285 **"ordered to wipe out my name":** Letter from Mark Popovsky to Alexander Chakovsky, chief editor of *Literaturnaya Gazeta*, January 17, 1988, 4078435, CUL.

286 **seeds from Vavilov's collection:** In an article for the Russian edition of *Newsweek*, the former director of the VIR, Viktor Dragavtsev, criticized his successor, Nikolai Dzyubenko, for not sending a duplicate of the entire Vavilov collection to Svalbard. Nikita Malsimov, "Kolos: Na Glinyanykh Nogakh," *Newsweek*, March 3, 2008.

286 **"a mythic resonance":** John Seabrook, "Sowing for Apocalypse," *New Yorker*, August 27, 2007.

288 **"The seeds age unevenly":** Borin, *Krutyye povoroty*, 17ff.

288 **"cannot be fathomed":** As reprinted in Borin, "Podvig 13 leningradtsev."

288 **"Imagine this scenario":** Borin, *Krutyye povoroty*, 18.

289 **Ivanov edited two editions:** "Remembering Vavilov," 1966, and "The Founder of VIR," 1968.

290 **a letter to his friend:** Courtesy of Uwe Hossfeld.

290 **"may prove of inestimable value":** As quoted in Pearce, "Great Seed Blitzkrieg," 41.

290 **claimed to have come from Lannach:** Ibid., 40.

290 **in exchange for a stateless passport:** Letter from Heinz Brücher to Herr Hofrat, November 7, 1948, Erich Tschermak-Seysenegg estate, box 10, AIOA.

291 **citing a lack of evidence:** Email from Arnold Steinbrecher to Uwe Hossfeld, June 13, 2007, courtesy of Uwe Hossfeld.

291 **"would have killed me, too":** Mat Youkee, "Who Killed the Nazi Botanist Trying to Wipe Out Cocaine?," OZY, January 26, 2018.

AFTERWORD

296 **on an earlier trip to China:** Leather pouch containing packets of seeds from the *Henriette*, catalog reference EXT 11/140, extracted from HCA 32/1048, TNA.

300 **clear the land for upscale dachas:** "The Battle to Save Russia's Pavlovsk Seed Bank," *Guardian*, September 20, 2010.

301 **"destroy the collection in three hours":** Nikita Malsimov, "Kolos: Na Glinyanykh Nogakh," *Newsweek*, March 3, 2008.

BIBLIOGRAPHY

ARCHIVAL SOURCES

AIOA, Österreichische Akademie der Wissenschaften.
BDC, Berlin Document Center.
CUL, Columbia University Libraries Special Collections.
EPE, Embryo Project Encyclopaedia (https://embryo.asu.edu).
NKVD, Investigation Files of the Soviet Secret Police (collected by Mark Popovsky).
TNA, The National Archives, Kew, Surrey.
TRS, The Royal Society.
VIR, Vavilov Institute Archives.

FILMS

Rod MacLeish. *The Hermitage: A Russian Odyssey*. Christian Science Publishing Society, 1994.
Zavtra ne umret nikogda: Zolotoy zeloniy zapas. Rossiya-Kultura, 2012–15.

UNPUBLISHED SOURCES

Bailey, Lane. "Protecting the Art of Leningrad: The Survival of the Hermitage Museum during the Great Patriotic War." Honors thesis. Ouachita Baptist University, 1997.
Bericht über das Sammelkommando. SS-Akte Brücher, Bl. 60/61. BDC. Translation courtesy of Dr. Anna Nyburg.
Biological Journals of the Linnean Society.
Ivanov, N. R. "Sokhraneniye Kollektsii Kul'turnykh Rasteniy V.I.R. v Leningrade v Gody Blokady." Unpublished memoir, undated. Courtesy of the Vavilov Institute.

Kameraz, Abram Yakovlevich. Unpublished autobiography. February 25, 1949. Courtesy of the Vavilov Institute.

The N. I. Vavilov Institute of Plant Industry, 1925–1975. VIR pamphlet, 1975.

Orders for the All-Union Plant Institute (various). Courtesy of the Vavilov Institute.

Savitsky, Helen. "My Remembrance of N. I. Vavilov." Unpublished manuscript.

Sigunov, Egor V., and Ivan A. Fokin. *Posledovateli Nikolaya Vavilova.* Self-published book, 2020.

Spisok spetsialistov, imeyushchikh vyssheye obrazovaniye, vybyvshikh s pred-priyatiya (uchrezhdeniya), October 1941–January 1942. Courtesy of the Vavilov Institute.

Staffing Report, June 1–December 31, 1941. Courtesy of the Vavilov Institute.

Staff profiles (various). Courtesy of the Vavilov Institute.

PUBLISHED PAPERS

Alexanyan, S. M., and V. I. Krivchenko. "Vavilov Institute Scientists Heroically Preserve World Plant Genetic Resources Collections during World War II Siege of Leningrad." *Diversity* 7, no. 4 (1991): 10ff.

Bidlack, Richard. *Workers at War: Factoryworkers and Labor Policy in the Siege of Leningrad.* No. 902. University of Pittsburgh Center for Russian and East European Studies, March 1991.

Borin, Alexander. "Podvig 13 leningradtsev. K 125-letiyu Nikolai Vavilov." *Novaya Gazeta,* November 23, 2012. Original interview 1976.

Elina, Olga. "From Russia with Seeds: The Story of the Savitskys, Plant Geneticists and Breeders." *Studies in the History of Biology* 6, no. 2 (2014).

Geyer, Georgie Anne. "Nine Hundred Days of Sacrifice." *International Wildlife,* January–February 1973.

Hossfeld, Uwe, and Carl-Gustaf Thornström. "Instant Appropriation—Heinz Brücher and the SS Botanical Collecting Commando to Russia, 1943." *Plant Genetic Resources Newsletter* 39 (2002).

———. "Retrospektive einer Botanikerflucht nach dem Ende des Zweiten Welt-krieges: Drei Briefe von Heinz Brücher an Erich Tschermak-Seysenegg (1948–1950)." *Haussknechtia* 10 (2004): 267–97.

Kolchinsky, Eduard. "Nikolai Vavilov in the Years of Stalin's 'Revolution from Above' (1929–1932)." *Centaurus* 56, no. 4 (November 2014): 330–58.

Kolchinsky, Eduard, Uwe Hossfeld, and Georgy S. Levit. "Russian Scientists and the Royal Society of London: 350 Years of Scientific Collaboration." *Notes and Records: The Royal Society of the History of Science* 72, no. 3 (September 2018).

Loskutov, I. G. "Wartime Activities of the Vavilov Institute." *Proceedings on Applied Botany, Genetics and Breeding* 182, no. 2 (2021): 151–62.

Pearce, Fred. "The Great Seed Blitzkrieg." *New Scientist* 197, no. 2638 (January 12, 2008): 38–41.

Prezent, I. I. "Pseudoscientific 'Theories in Genetics." *Vernalization*, no. 2 (1939).

Roll-Hansen, Nils. "Wishful Science: The Persistence of T. D. Lysenko's Agrobiology in the Politics of Science." *Osiris* 23, no. 1 (2008): 166–88.

Sokolov. "'Tyoplaya vanna dlya begemota: zoosad v gody voiny," *Rodina*, no. 1, 2003.

BOOKS

Adamovich, Ales, and Daniil Granin. *A Book of the Blockade*. Moscow: Raduga, 1983.

———. *Leningrad under Siege*. Yorkshire, UK: Pen & Sword Military, 2007.

Akhmatova, Anna Andreevna. *Selected Poems*. London: Penguin, 1988.

Alexander, Ralph B. *Science under Attack: The Age of Unreason*. New York: Algora Publishing, 2018.

Alexanian, S. M., ed. *N. I. Vavilov All-Russian Scientific Research Institute of Plant Industry*. St. Petersburg: VIR, 1994.

Anderson, M. T. *Symphony for the City of the Dead: Dmitri Shostakovich and the Siege of Leningrad*. Somerville, MA: Candlewick Press, 2015.

Bakhteev, F. Kh. *Nikolay Ivanovich Vavilov, 1887–1943*. Novosibirsk, Russia: Nauka Press, 1988.

Barber, John, and Andrei Dzeniskevich. *Life and Death in Besieged Leningrad, 1941–44*. New York: Palgrave Macmillan, 2004.

Barskova, Polina. *Living Pictures*. London: Pushkin Press, 2020.

Beevor, Anthony. *Stalingrad*. London: Penguin, 1999.

Bergholz, Olga. *Govirot Leningrad*. Moscow: AST Publishing, 2015.

Bidlack, Richard, and Nikita Lomagin. *The Leningrad Blockade, 1941–1944: A New Documentary History from the Soviet Archives*. Translated by Marian Schwartz. New Haven, CT: Yale University Press, 2012.

Birstein, Vadim J. *The Perversion of Knowledge*. Boulder, CO: Westview, 2001.

Boldyrev, Aleksandr. *Osadnaya zapis: Blokadniy dnevnik*. St. Petersburg, 1998.

Borin, Alexander. *Krutyye povoroty*. Moscow: Profizdat, 1982.

Brücher, Heinz. *Useful Plants of Neotropical Origin: And Their Wild Relatives*. New York: Springer, 1989.

Burdick, Charles, and Hans-Adolf Jacobsen, eds. *The Halder War Diary, 1939–1942*. Novato, CA: Presidio, 1988.

Cameron, Norman, and R. H. Steven. *Hitler's Table Talk, 1941–1944: His Private Conversations*. New York: Enigma Books, 2000.

Conquest, Robert. *The Great Terror: A Reassessment*. New York: Oxford University Press, 1990.

Deichmann, Ute. *Biologists under Hitler*. London: Harvard University Press, 1996.

Del Curto, Mario. *Seeds of the Earth: The Vavilov Institute*. Arles, France: Actes Sud, 2017.

Dimbleby, Jonathan. *Barbarossa*. London: Viking, 2021.

Dragavstev, V. A., ed. *Soratniki Nikolaya Ivanovicha Vavilova*. St. Petersburg: VIR, 1994.

Dzeniskevich, Andreii. *Leningrad v osade: Sbornik dokumentov*. St. Petersburg: Liki Rossii, 1995.

Dzybenko, N. I., E. I. Gaevskaya, M. A. Vishnyakova, I. G. Loskutov, S. N. Kutuzova, L. Y. Shipilina, I. V. Kotelkina, and E. A. Sokolova, eds. *Soratniki Nikolaya Ivanovicha Vavilova*, 2nd rev. ed. St. Petersburg: VIR, 2017.

Fedotova, A. A., and N. P. Goncharov. *Byuro Po Prikladnoy Botanike v Gody Pervoy Mirovoy Voyny*. St. Petersburg: Nestor-Istoria, 2014.

Gerhard, Gesine. *Nazi Hunger Politics: A History of Food in the Third Reich*. London: Rowman & Littlefield, 2015.

Ginzburg, Lydia. *Notes from the Blockade*. London: Vintage, 2016.

Glantz, David M. *The Siege of Leningrad, 1941–44*. London: Brown, 2001.

Golubev, G. *Nikolai Vavilov: The Great Sower: Pages from the Life of the Scientist*. Translated by Vadim Sternik. Moscow: Mir Publishers, 1987.

Graham, Loren. *Lysenko's Ghost*. Cambridge, MA: Harvard University Press, 2016.

Hawks, John Gregory, ed. *N. I. Vavilov Centenary Symposium*. London: Academic Press, 1990.

Höhne, Heinz. *Canaris: Hitler's Master Spy*. New York: Doubleday, 1979.

Inber, Vera. *Leningrad Diary*. Translated by Serge W. Wolff and Rachel Grieve. New York: St. Martin's, 1971.

Ings, Simon. *Stalin and the Scientists*. London: Faber & Faber, 2016.

International Military Tribunal. Trials of war criminals before the Nuernberg Military Tribunals under Control Council law no. 10 Nuernberg, October–April 1949. Washington: U.S. G.P.O., 1953.

Irving, David. *Hitler's War*. London: Viking, 1977

Jones, Michael. *Leningrad: State of Siege*. New York: Basic, 2008.

———. *The Retreat: Hitler's First Defeat*. London: John Murray, 2009.

Ivanov, N. R., V. S. Lekhnovich, K. A. Nikitin. *V Osazhdennom Leningrade*. Leningrad: Lenizdat, 1969.

Karner, Stefan, Heide Gsell, and Philipp Lesiak. *Lannach Castle, 1938–1949*. Graz, Austria: Leykam, 2007.

Kay, Alex. *Exploitation, Resettlement, Mass Murder: Political and Economic Planning for German Occupation Policy in the Soviet Union: 1940–41*. New York: Berghahn Books, 2006.

Kochina, Yelena. *Blockade Diary*. New York: Ardis, 1990.

Kohzina, Elena. *Through the Burning Steppe: A Wartime Memoir*. London: Duckworth, 2000.

Kurowski, Franz. *The Brandenburger Commandos*. Translated by David Johnston. Mechanicsburg, PA: Stackpole Military History Series, 2005.

Lazarev, Dmitri. *Vospominaniya o blokade*. St. Petersburg, 2000.

Likhachev, Dmitri S. *Reflections on the Russian Soul: A Memoir*. Budapest: CEU Press, 2000.

Loskutov, Igor. G. *Vavilov and His Institute*. Rome: IPGRI, 1999.

Lomagin, Nikita. *V Tiskakh Goloda: Blokada Leningrada v Dokumentakh Nemetskikh Spetssluzhb i NKVD*. St. Petersburg: SPB, 2000.

Madajczyk, Czeslaw. *Vom Generalplan Ost zum Generalsiedlungsplan*. Berlin, New York: K. G. Saur, 1994.

Massie, Suzanne. *Pavlovsk: The Life of a Russian Palace*. London: Little, Brown, 1990.

Mawdsley, Evan. *Thunder in the East: The Nazi-Soviet War, 1941–1945*. London: Bloomsbury, 2007.

Medvedev, Z. A. *The Rise and Fall of T. D. Lysenko*. Translated by M. I. Lerner. New York: Columbia University Press, 1969.

———. *The Unknown Stalin: His Life, Death and Legacy*. London: I. B. Tauris, 2005.

———. *Vzlet i Padenie Lysenko: Istoriya Biologicheskoi Diskussii v USSR (1929–1966)*. Moscow: Kniga, 1993.

Meyer, Karl, and Shareen Brysac. *Tournament of Shadows*. London: Little, Brown, 2001.

Millar, David, Ian Millar, John Millar, and Margaret Millar. *Cambridge Dictionary of Scientists*. 2nd ed. Cambridge: Cambridge University Press, 2002.

Mukhina, Elena. *The Diary of Lena Mukhina*. London: Macmillan, 2015.

Nabhan, Gary Paul. *Where Our Food Comes From*. Washington, DC: Island Press, 2009.

Nabokov, Vladimir. *Speak, Memory: An Autobiography Revisited*. London: Penguin Modern Classics, 1966.

Nikitin, Vladimir. *The Unyielding City: Blockade, 1941–44*. St. Petersburg: Limbus Press, 2019.

Nordqvist, Jens. *Den Stora Frostolden*. Lund, Sweden: Historika Media, 2020.

Overy, Richard. *Russia's War: A History of the Soviet War Effort, 1941–45*. New York: Penguin, 1997.

Papernaya, N. N. *Podvig khraniteley: Kto i kak spasal muzeyniye tsennosti v blokadnom Leningrade*. Leningrad, 1969.

Paterson, Lawrence. *Hitler's Brandenburgers*. London: Greenhill Books, 2018.

Pavlov, Dimitri. *Leningrad, 1941: The Blockade*. Translated by John Adams. Chicago: University of Chicago Press, 1965.

Peri, Alexis. *The War Within: Diaries from the Siege of Leningrad*. (Boston: Harvard University Press, 2017.

Petrovskaya Wayne, Kyra. *Shurik: A WWII Saga of the Siege of Leningrad*. New York: Lyons, 2000.

Pimenova, Victoria, ed. *Deti voiny: Narodnaya kaniga pamyati* Moscow: AST Publishing, 2015.

Popovsky, Mark. *The Vavilov Affair*. Hamden, CT: Archon Books, 1984.

Pringle, Heather. *The Master Plan: Himmler's Scholars and the Holocaust*. New York: Hyperion, 2006.

Pringle, Peter. *The Murder of Nikolai Vavilov*. New York: Simon & Schuster, 2008.

Reid, Anna. *Leningrad: Tragedy of a City under Siege, 1941–44*. London: Bloomsbury, 2011.

Robinson, Daniel F. *Confronting Biopiracy: Challenges, Cases, and International Debates*. London: Routledge, 2010.

Rokityansky, Yakov, Yuri Vavilov, and V. Goncharov. *Sud Palacha: Nikolai Vavilov v Zastenkakh NKVD*. Moscow: Academia Moskva, 1999.

Roll-Hansen, Nils. *The Lysenko Effect: The Politics of Science*. New York: Humanity Books, 2005.

Salisbury, Harrison. *The 900 Days*. London: Pan Books. 2000.

Senin, Viktor. *List'ya vyrastut vnov'*. Leningrad: Lenizdat, 1979.

Shentalinsky, Vitaly. *The KGB's Literary Archive*. London: Harvill, 1995.

Shirer, William L. *The Rise and Fall of the Third Reich*. New York: Simon & Schuster, 1960.

Shukman, Harold. *Stalin's Generals*. London: Weidenfeld & Nicolson, 1993.

Simmons, Cynthia, and Nina Perlina, eds. *Writing the Siege of Leningrad: Women's Diaries, Memoirs and Documentary Prose*. Pittsburgh: University of Pittsburgh Press, 2002.

Skrjabina, Elena. *Siege and Survival*. Translated by Norman Luxenburg. Livingston, NJ: Transaction, 1997.

Sorokin, Pitirim A. *Leaves from a Russian Diary—and Thirty Years After*. Boston: Beacon Press, 1950.

Soyfer, Valerey. *Lysenko and the Tragedy of Soviet Science*. Translated by Leo and Rebecca Gruliow. New Brunswick, NJ: Rutgers University Press, 1994.

Spaeter, Helmuth. *Die Brandenburger: Eine deutsche Kommandotruppe*. Munich: Walter Angerer, 1978.

Suldin, Andrey. *Blokada Leningrada. 872 dnya i nochi. Polnaya khronika*. Moscow: AST, 2015.

Thubron, Colin. *In Siberia*. London: Vintage, 1999.

Tikhonov, Nikolai. *The Defence of Leningrad*. London: Hutchinson, 1943.

Tooze, Adam. *The Wages of Destruction: The Making and Breaking of the Nazi Economy*. London: Penguin, 2006.

Varshavsky, Sergei, and Boris Rest. *Saved for Humanity: The Hermitage during the Siege of Leningrad, 1941–1944*. London: Aurora, 1985.

Vavilov, N. I. *Five Continents*. Translated by Doris Löve. Rome: International Plant Genetic Resources Institute, 1997.

Vavilov, S. I. *Ocherki i Vospominaniya*. Moscow: Izdatelstvo Nauka, 1979.

Veselov, Aleksandr, ed. *The Unyielding City: Blockade, 1941–1944*. St. Petersburg: Limbus Press, 2019.

Volkogonov, Dmitrii Antonovich. *Stalin: Triumph and Tragedy*. New York: Weidenfeld & Nicolson, 1991.

Volkov, Solomon. *St. Petersburg: A Cultural History*. New York: Free Press Paperbacks, 1997.

Zherve, A. V., A. Karpenko, A. N. Kolyakin, and Iu. B. Demidenko. *Trudy Gosudarstvennogo Muzeya Istorii Sankt-Peterburga*. St. Petersburg: GMI SPb, 2012.

Zhukov, Georgi. *The Memoirs of Marshal Zhukov*. London: Delacorte Press, 1971.

Zyatkov, Nikolai, Evgeny Faktorovich, Tatyana Kuznetsova, and Polina Ivanushkina, eds. *Detskaya kniga voiny*. Moscow: Argumenty i Fakty, 2016.

INDEX

NOTE: Page references in *italics* refer to photos. Page references in **bold** refer to Plant Institute staff roll call information.